Lecture Notes in Mathematics

Edited by A. Dold and B. Eckmann

440

Ronald K. Getoor

Markov Processes: Ray Processes and Right Processes

Springer-Verlag
Berlin · Heidelberg · New York 1975

Prof. Ronald K. Getoor
Department of Mathematics
University of California
San Diego
P.O. Box 109
La Jolla, CA 92037/USA

Library of Congress Cataloging in Publication Data

Getoor, Ronald Kay, 1929-
Markov processes, ray processes and right processes.

(Lecture notes in mathematics : 440)
Bibliography: p.
Includes indexes.
1. Markov processes. I. Title. II. Series:
Lecture notes in mathematics (Berlin) ; 440.
QA3.L28 no. 440 [QA274.7] 510'.8s [519.2'33]
75-6610

AMS Subject Classifications (1970): 60JXX, 60J25, 60J40, 60J45, 60J50

ISBN 3-540-07140-7 Springer-Verlag Berlin · Heidelberg · New York
ISBN 0-387-07140-7 Springer-Verlag New York · Heidelberg · Berlin

Offsetdruck: Julius Beltz, Hemsbach/Bergstr.

PREFACE

The purpose of these lectures is to develop the basic properties of Ray processes and their applications to processes satisfying the "hypothèses droites" of Meyer following the fundamental paper [16] by Meyer and Walsh. Sections 2 through 7 of these lectures discuss the basic results on Ray processes and, in outline, follow quite closely the presentation in Section 1 of [16]. However, we go into more detail than in [16] and, in particular, we give complete proofs of the facts needed about resolvents and semigroups in Sections 2 and 3.

Beginning in Section 9 we give the basic applications of Ray processes to "right processes" again following [16] in outline. However, we depart from Meyer and Walsh in two important matters. Firstly we assume only that the state space E is a U-space, that is, a universally measurable subspace of a compact metric space; whereas Meyer and Walsh assume that E is Lusinien, that is, a Borel subspace of a compact metric space. Secondly we do not assume that the excessive functions are nearly Borel. We assume only that they are right continuous along the trajectories of the process. This change in the "hypothèses droites" sometimes requires a modification in the proofs of the basic results. Thus the statements of the theorems in Sections 10 through 13 are the same as in Meyer and Walsh, but often the proofs are somewhat different. The basic definitions and elementary properties of right processes are given in Section 9. The Ray-Knight compactification is presented in Section 10, while in Section 11 it is shown that the results on Ray processes developed in Sections 5 through 7 actually hold when properly interpreted for right processes. This is the most important part of these lectures. Section 12 contains Shih's theorem which was the catalyst for the renewed interest in the Ray-Knight construction. It also contains the pleasing result that the excessive functions are nearly Borel after all, but in the Ray topology. Section 13 discusses the relationships among right processes, Hunt processes, and standard processes. Finally in Sections 14 and 15 we investigate to what extent the preceding constructions are unique.

These results are taken from my joint paper with M. J. Sharpe [6].

The reader of these lectures should be familiar with the general theory of processes as set forth in the recent book of Dellacherie [3]. He should also have some acquaintance with the strong Markov property and the construction of Markov processes from transition functions as presented in Sections 8 and 9 of Chapter I of Blumenthal and Getoor [1], or in Chapters XII and XIII of Meyer [9]. However, an extensive knowledge of Markov processes is not required.

We have made no attempt to assign credits for the results discussed here. Suffice it to say that all of the basic ideas come from Knight [7], Meyer [9], Ray [14], Shih [15], and Meyer and Walsh [16]. Our only contribution is the relaxation of the assumptions on the state space and the excessive functions and the results in Sections 14 and 15 as mentioned before. See Meyer [12] in this connection also.

I would like to thank M. J. Sharpe for many helpful discussions and suggestions during the writing of Sections 9 through 13, and to reiterate that Sections 14 and 15 are based on our joint paper [6]. C. Gzyl and P. Protter read most of the manuscript and made innumerable suggestions for improving the exposition. C. Gzyl also helped with the proofreading of the final typescript. L. Smith and A. Whiteman typed the preliminary and final versions respectively. Their superb skill greatly eased my work in preparing the manuscript. Finally I would like to thank the National Science Foundation for financial support during part of the writing under NSF Grant GP-41707X.

R. K. Getoor
La Jolla, California
November, 1974

CONTENTS

1. PRELIMINARIES

The reader of these lectures is assumed to be familiar with the general theory of processes as presented in the recent book of Dellacherie [3]. We shall refer constantly to [3] and adopt the following convention: A reference to D-II-19 will refer to item 19 of Chapter II in Dellacherie [3]. In addition, the reader is assumed to have some acquaintance with the strong Markov property and the construction of Markov processes from transition functions as set forth, for example, in Sections 8 and 9 of Chapter I of Blumenthal and Getoor [1]. However, an extensive knowledge of Markov processes is not assumed. A reference to BG-III-(4.19) will refer to item (4.19) of Chapter III in Blumenthal and Getoor [1].

In general, our notation will be the same as that in Blumenthal and Getoor. In particular it is assumed that the reader is familiar with the notation established in Section 1 of Chapter 0 in BG. For example, if $(E, \underline{\underline{E}})$ is a measurable space and f a numerical function on E, we write $f \in \underline{\underline{E}}$ to indicate that f is $\underline{\underline{E}}$ measurable. We let $b\underline{\underline{E}}$ denote the bounded real valued $\underline{\underline{E}}$ measurable functions on E and $b\underline{\underline{E}}^+$ (or sometimes $b\underline{\underline{E}}_+$) the positive functions in $b\underline{\underline{E}}$. Similarly $f \in \underline{\underline{E}}^+$ (or $f \in \underline{\underline{E}}_+$) means that f is a positive $\underline{\underline{E}}$ measurable function on E. In accordance with modern usage positive means nonnegative. A numerical function f on E is strictly positive if $f(x) > 0$ for all x in E. We let $\underline{\underline{E}}^*$ denote the σ-algebra of universally measurable sets over $(E, \underline{\underline{E}})$. By a measure on a measurable space $(E, \underline{\underline{E}})$ we shall always mean a positive measure.

If $(E, \underline{\underline{E}})$ and $(F, \underline{\underline{F}})$ are measurable spaces, then we write $f \in \underline{\underline{E}}|\underline{\underline{F}}$ or say that f is $\underline{\underline{E}}|\underline{\underline{F}}$ measurable whenever f is a measurable mapping from $(E, \underline{\underline{E}})$ to $(F, \underline{\underline{F}})$; that is, $f\colon E \to F$ and $f^{-1}(B) \in \underline{\underline{E}}$ for all $B \in \underline{\underline{F}}$. If μ is a measure on $(E, \underline{\underline{E}})$ and $f \in \underline{\underline{E}}|\underline{\underline{F}}$ we write $\nu = f(\mu)$ for the measure ν defined on $(F, \underline{\underline{F}})$ by $\nu(B) = \mu[f^{-1}(B)]$ for all $B \in \underline{\underline{F}}$. The measure $\nu = f(\mu)$ is called the image of μ under f. (In BG this measure was denoted by μf^{-1} rather than the more standard $f(\mu)$.)

If E is a topological space, then the Borel subsets of E are the elements of the smallest σ-algebra containing all of the open subsets of E. We shall denote this σ-algebra by $\underline{\underline{E}}$ or, sometimes, by $\underline{\underline{B}}(E)$. The universally measurable subsets of a topological space are the elements of $\underline{\underline{E}}^* = \underline{\underline{B}}^*(E)$ where $\underline{\underline{E}} = \underline{\underline{B}}(E)$ is the σ-algebra of Borel subsets of E. We let $\mathbb{R}$ denote the real numbers, $\mathbb{R}^+ = [0, \infty)$ the nonnegative reals, and $\mathbb{R}^{++} = (0, \infty)$ the strictly positive reals. Then $\underline{\underline{\mathbb{R}}}$, $\underline{\underline{\mathbb{R}}}^+$, and $\underline{\underline{\mathbb{R}}}^{++}$ are the σ-algebras of Borel subsets of $\mathbb{R}$, $\mathbb{R}^+$, and $\mathbb{R}^{++}$ respectively.

If $(E, \underline{\underline{E}})$ and $(F, \underline{\underline{F}})$ are measurable spaces, then a <u>kernel</u> K from $(F, \underline{\underline{F}})$ to $(E, \underline{\underline{E}})$ is a positive function $K(x, A)$ defined for $x \in F$ and $A \in \underline{\underline{E}}$ such that $x \to K(x, A)$ is $\underline{\underline{F}}$ measurable for each $A \in \underline{\underline{E}}$ and $A \to K(x, A)$ is a measure on $(E, \underline{\underline{E}})$ for each $x \in F$. The kernel K is finite if $K(x, E) < \infty$ for all x and bounded if $\sup\{K(x, E): x \in F\} < \infty$. If $K(x, E) = 1$ for all $x \in F$, then K is a Markov kernel; if $K(x, E) \le 1$ for all $x \in F$, then K is a sub-Markov kernel. If K is a bounded kernel from $(F, \underline{\underline{F}})$ to $(E, \underline{\underline{E}})$, then

(1.1) $$f \to Kf; \text{ where } Kf(x) = K(x, f) = \int K(x, dy)\, f(y)$$

defines a bounded, linear, positive map from $b\underline{\underline{E}}$ to $b\underline{\underline{F}}$ such that

(1.2) $$(f_n) \subset b\underline{\underline{E}}^+;\ 0 \le f_n \uparrow f \in b\underline{\underline{E}} \Rightarrow Kf_n \uparrow Kf .$$

Conversely any bounded, linear, positive map from $b\underline{\underline{E}}$ to $b\underline{\underline{F}}$ satisfying (1.2) is given by a bounded kernel K from $(F, \underline{\underline{F}})$ to $(E, \underline{\underline{E}})$ as in (1.1).

If $(E, \underline{\underline{E}})$, $(F, \underline{\underline{F}})$ and $(G, \underline{\underline{G}})$ are measurable spaces and K is a kernel from $(F, \underline{\underline{F}})$ to $(E, \underline{\underline{E}})$ and L is a kernel from $(G, \underline{\underline{G}})$ to $(F, \underline{\underline{F}})$, then the composition of K and L, LK is a kernel from $(G, \underline{\underline{G}})$ to $(E, \underline{\underline{E}})$ defined by

(1.3) $$LK(x, A) = \int L(x, dy)\, K(y, A)$$

for $x \in G$ and $A \in \underline{\underline{E}}$ where the integration in (1.3) is over F. If K and L are bounded (resp. Markov, sub-Markov), then LK is bounded (resp. Markov, sub-Markov). By a kernel on $(E, \underline{\underline{E}})$ we shall mean a kernel from $(E, \underline{\underline{E}})$ to $(E, \underline{\underline{E}})$.

2. RESOLVENTS

Throughout this section $(E, \underline{\underline{E}})$ is a fixed measurable space.

(2.1) DEFINITION. *A family* $(U^\alpha)_{\alpha>0}$ *of kernels on* $(E, \underline{\underline{E}})$ *is a (sub-Markov) resolvent provided:*

(i) $\alpha U^\alpha 1 \le 1$ *for* $\alpha > 0$.

(ii) $U^\alpha - U^\beta = (\beta - \alpha) U^\alpha U^\beta$ *for* $\alpha, \beta > 0$.

It is Markov if $\alpha U^\alpha 1 = 1$ *for* $\alpha > 0$.

In general we shall omit the qualifying phrase "sub-Markov"; that is a resolvent will always mean a sub-Markov resolvent. Since $U^\alpha(x, E) \le \alpha^{-1}$ for all x, each U^α is a bounded kernel on $(E, \underline{\underline{E}})$ and so there is no difficulty with the subtraction in (ii). The relationship (ii) is called the resolvent equation. Note that it is αU^α that is a sub-Markov kernel and not U^α itself.

(2.2) REMARKS. (a) It is immediate from (ii) that $U^\alpha U^\beta = U^\beta U^\alpha$ for $\alpha, \beta > 0$, and that $\alpha \to U^\alpha(x, \cdot)$ is decreasing and continuous on $(0, \infty)$. Consequently $U(x, \cdot) = \lim_{\alpha \to 0} U^\alpha(x, \cdot)$ defines a kernel on $(E, \underline{\underline{E}})$, but, in general, $U(x, \cdot)$ need not be finite (or even σ-finite).

(b) If $\beta > 0$ and we define $V^\alpha = U^{\alpha+\beta}$ for $\alpha > 0$, then it is immediate that $(V^\alpha)_{\alpha>0}$ is a resolvent and that $V = \lim_{\alpha \to 0} V^\alpha = U^\beta$ is a bounded kernel.

(c) If $f \in \underline{\underline{E}}^+$ and $\beta > \alpha$, then the resolvent equation implies that $U^\alpha f = U^\beta f + (\beta - \alpha) U^\alpha U^\beta f$, even though $U^\alpha f - U^\beta f$ is undefined in general.

(d) It is immediate from (c) that if $f \in \underline{\underline{E}}^+$, $\alpha \to U^\alpha f$ is decreasing on $(0, \infty)$.

(e) If $f \in b\underline{\underline{E}}$, then from the resolvent equation and (2.1-i)

$$|U^\alpha f(x) - U^\beta f(x)| \le \frac{|\beta - \alpha|}{\alpha\beta} \|f\|$$

where $\|f\| = \sup\{|f(x)| : x \in E\}$. Consequently $\alpha \to U^\alpha f(x)$ is continuous uniformly in x on each interval $[\alpha_0, \infty)$, $\alpha_0 > 0$.

(f) If $(U^\alpha)_{\alpha>0}$ is a resolvent on $(E, \underline{\underline{E}})$, then it is easy to check that $(U^\alpha)_{\alpha>0}$ is also a resolvent on $(E, \underline{\underline{E}}^*)$. This amounts to checking that $x \to U^\alpha(x, A)$ is $\underline{\underline{E}}^*$ measurable whenever $A \in \underline{\underline{E}}^*$.

(2.3) DEFINITION. Let $f \in \underline{\underline{E}}^+$ and $\alpha \geq 0$. Then f is α-supermedian provided $\beta U^{\alpha+\beta} f \leq f$ for all $\beta > 0$. If, in addition, as $\beta \to \infty$, $\beta U^{\alpha+\beta} f \to f$ pointwise, then f is α-excessive. We let $\mathcal{S}^\alpha$ (resp. $\mathcal{E}^\alpha$) denote the set of all α-supermedian (resp. α-excessive) functions. We write $\mathcal{S} = \mathcal{S}^0$ and $\mathcal{E} = \mathcal{E}^0$, and say that f is supermedian (resp. excessive) rather than 0-supermedian (resp. 0-excessive).

The next proposition collects a number of elementary properties of supermedian and excessive functions.

(2.4) PROPOSITION. (i) $\mathcal{S}^\alpha$ and $\mathcal{E}^\alpha$ are convex cones; $\mathcal{S}^\alpha$ is closed under pointwise infima, i.e., if $f, g \in \mathcal{S}^\alpha$ then $f \wedge g = \min(f, g) \in \mathcal{S}^\alpha$; $1 \in \mathcal{S}^\alpha$.

(ii) If $f \in \mathcal{S}^\alpha$, then $\beta \to \beta U^{\alpha+\beta} f(x)$ is increasing for each x.

(iii) If (f_n) is an increasing sequence in $\mathcal{S}^\alpha$ (resp. $\mathcal{E}^\alpha$), then $f = \lim f_n$ is in $\mathcal{S}^\alpha$ (resp. $\mathcal{E}^\alpha$).

(iv) $f \in \mathcal{S}^\alpha$ if and only if $f \in \mathcal{S}^\beta$ for all $\beta > \alpha$. (The corresponding statement for $\mathcal{E}^\alpha$ is also true - see (2.8).)

(v) If $f \in \underline{\underline{E}}^+$, then $U^\alpha f \in \mathcal{E}^\alpha$.

PROOF. (i) This is elementary and left to the reader.

(ii) Let $f \in \mathcal{S}^\alpha$ and fix $0 < \beta < \gamma$. Then $\beta U^{\alpha+\beta} f \leq f$. Applying $(\gamma - \beta) U^{\alpha+\gamma}$ to this inequality gives

$$\beta(\gamma-\beta) U^{\alpha+\gamma} U^{\alpha+\beta} f \leq (\gamma-\beta) U^{\alpha+\gamma} f ,$$

and adding $\beta U^{\alpha+\gamma} f$ to both sides we find

$$(2.5) \quad \beta[U^{\alpha+\gamma} f + (\gamma-\beta) U^{\alpha+\gamma} U^{\alpha+\beta} f] \leq \beta U^{\alpha+\gamma} f + (\gamma-\beta) U^{\alpha+\gamma} f = \gamma U^{\alpha+\gamma} f .$$

This last equality is clear at each x such that $U^{\alpha+\gamma} f(x) < \infty$, but it is also clear if $U^{\alpha+\gamma} f(x) = \infty$ since $\gamma > \beta$. By the resolvent equation, or more precisely (2.2-c), the left side of (2.5) reduces to $\beta U^{\alpha+\beta} f$, proving (ii).

(iii) Suppose $0 \le f_n \uparrow f$. If $(f_n) \subset \mathcal{S}^\alpha$, then $\beta U^{\alpha+\beta} f_n \le f_n$. Letting $n \to \infty$ the monotone convergence theorem gives $\beta U^{\alpha+\beta} f \le f$, and so $f \in \mathcal{S}^\alpha$. If each $f_n \in \mathcal{E}^\alpha \subset \mathcal{S}^\alpha$, then (ii) and the definition of $\mathcal{E}^\alpha$ imply that $\beta U^{\alpha+\beta} f_n \uparrow f_n$ as $\beta \to \infty$. Therefore $\beta U^{\alpha+\beta} f_n$ increases both with n and β, and so

$$\lim_\beta \beta U^{\alpha+\beta} f = \lim_\beta \lim_n \beta U^{\alpha+\beta} f_n = \lim_n \lim_\beta \beta U^{\alpha+\beta} f_n = f .$$

Hence $f \in \mathcal{E}^\alpha$.

(iv) In view of (i) and (iii), $f \in \mathcal{S}^\alpha$ if and only $f \wedge n \in \mathcal{S}^\alpha$ for all $n \ge 1$. Consequently in proving (iv) we may assume f bounded. If $\beta > \alpha$ and $f \in \mathcal{S}^\alpha$, then for $\gamma > 0$, $f \ge \gamma U^{\alpha+\gamma} f \ge \gamma U^{\beta+\gamma} f$ and so $f \in \mathcal{S}^\beta$. Conversely if $\gamma U^{\beta+\gamma} f \le f$ for all $\beta > \alpha$, then letting β decrease to α and using (2.2-e) yields $\gamma U^{\alpha+\gamma} f \le f$.

(v) By (iii) it suffices to show $U^\alpha f \in \mathcal{E}^\alpha$ when $f \in b\underline{\underline{E}}^+$. But $\gamma U^{\alpha+\gamma} U^\alpha f = U^\alpha f - U^{\alpha+\gamma} f \le U^\alpha f$ and $\| U^{\alpha+\gamma} f \| \le (\alpha+\gamma)^{-1} \| f \|$ which approaches zero as $\gamma \to \infty$. (Here, as before, $\| \cdot \|$ stands for the sup norm.) Thus $U^\alpha f \in \mathcal{E}^\alpha$.

(2.6) PROPOSITION. Let $f \in \mathcal{S}^\alpha$. Then $\hat{f} = \lim_{\beta \to \infty} \beta U^{\alpha+\beta} f$ exists and $\hat{f} \in \mathcal{E}^\alpha$. Also $\hat{f} \le f$ and $U^\beta f = U^\beta \hat{f}$ for all $\beta > 0$. The function $\hat{f}$ is called the (α-excessive) regularization of f and is the largest α-excessive function dominated by f.

PROOF. We know that $\beta \to \beta U^{\alpha+\beta} f$ is increasing and dominated by f. Consequently $\hat{f}$ exists, $\hat{f} \le f$, and $\hat{f} \in \underline{\underline{E}}^+$. If $g \in \mathcal{E}^\alpha$ and $g \le f$, then $\beta U^{\alpha+\beta} g \le \beta U^{\alpha+\beta} f$, and letting $\beta \to \infty$ we see that $g \le \hat{f}$. Thus $\hat{f}$ dominates any α-excessive function which is dominated by f. Next we shall show that $\hat{f} \in \mathcal{E}^\alpha$. To this end suppose first that f is bounded. By the monotone convergence theorem $U^\beta \hat{f} = \lim_{\gamma \to \infty} U^\beta \gamma U^{\alpha+\gamma} f$ for all $\beta > 0$. But $U^\beta f - U^{\alpha+\gamma} f = (\alpha+\gamma-\beta) U^\beta U^{\alpha+\gamma} f$, and so

$$\gamma U^\beta U^{\alpha+\gamma} f = U^\beta f - U^{\alpha+\gamma} f + (\beta-\alpha) U^\beta U^{\alpha+\gamma} f .$$

Consequently $\gamma U^\beta U^{\alpha+\gamma} f \to U^\beta f$ in sup norm as $\gamma \to \infty$. Therefore $U^\beta \hat{f} = U^\beta f$ for all $\beta > 0$. Thus $\beta U^{\alpha+\beta} \hat{f} = \beta U^{\alpha+\beta} f \uparrow \hat{f}$ as $\beta \to \infty$ and so $\hat{f} \in \mathcal{E}^\alpha$.

For the general case let $f_n = f \wedge n \in \mathcal{S}^\alpha$. Then $\beta U^{\alpha+\beta} f_n$ increases both with n and β. Therefore

$$\hat{f} = \lim_{\beta} \beta U^{\alpha+\beta} f = \lim_{n} \lim_{\beta} \beta U^{\alpha+\beta} f_n = \lim_{n} \hat{f}_n .$$

Clearly $(\hat{f}_n)$ is increasing and since each $\hat{f}_n \in \mathcal{E}^\alpha$, it follows from (2.4-iii) that $\hat{f} \in \mathcal{E}^\alpha$. Finally we know that for each $\beta > 0$ and n, $U^\beta \hat{f}_n = U^\beta f_n$ and letting $n \to \infty$ we obtain $U^\beta \hat{f} = U^\beta f$, completing the proof of (2.6).

(2.7) REMARK. If $f \in \mathcal{E}^\alpha$, we do not know that $f \wedge n \in \mathcal{E}^\alpha$. Nevertheless, there exists a sequence (f_n) of bounded α-excessive functions with $f_n \uparrow f$. To see this let $f_n = f \wedge n \in \mathcal{S}^\alpha$. Then $\hat{f}_n \le f_n \le n$, $\hat{f}_n \in \mathcal{E}^\alpha$, and as in the proof of (2.6), $\hat{f}_n \uparrow \hat{f} = f$ since $f \in \mathcal{E}^\alpha$.

(2.8) COROLLARY. $f \in \mathcal{E}^\alpha$ _if and only if_ $f \in \mathcal{E}^\beta$ _for all_ $\beta > \alpha$.

PROOF. Suppose $f \in \mathcal{E}^\beta$ for all $\beta > \alpha$. Then $f \in \mathcal{S}^\alpha$ by (2.4-iv). Therefore $\gamma U^{\beta+\gamma} f \le \gamma U^{\alpha+\gamma} f \le f$ if $\beta > \alpha$, and since $\gamma U^{\beta+\gamma} f \to f$ as $\gamma \to \infty$ it follows that $f \in \mathcal{E}^\alpha$. For the converse fix $\beta > \alpha$. Then using (2.7) it suffices to show that if f is bounded and $f \in \mathcal{E}^\alpha$, then $f \in \mathcal{E}^\beta$. We know by (2.4-iv) that $f \in \mathcal{S}^\beta$. Now

$$\gamma U^{\alpha+\gamma} f = \gamma U^{\beta+\gamma} f + \gamma(\beta - \alpha) U^{\alpha+\gamma} U^{\beta+\gamma} f ,$$

and $\gamma U^{\alpha+\gamma} f \to f$ while

$$\| \gamma(\beta - \alpha) U^{\alpha+\gamma} U^{\beta+\gamma} f \| \le \frac{\gamma(\beta - \alpha)}{(\alpha + \gamma)(\beta + \gamma)} \| f \| \to 0$$

as $\gamma \to \infty$. Therefore $\gamma U^{\beta+\gamma} f \to f$, and so $f \in \mathcal{E}^\beta$.

(2.9) PROPOSITION. _If_ $f \in \mathcal{E}^\alpha$ _with_ $\alpha > 0$, _then there exists a sequence_ $(f_n) \subset b\underline{\underline{E}}^+$ _with_ $U^\alpha f_n \uparrow f$.

PROOF. Let (g_n) be a sequence of bounded functions in $\mathcal{E}^\alpha$ with $g_n \uparrow f$. Then $\beta U^{\alpha+\beta} g_n$ increases with both β and n, and increases to f as $\beta \to \infty$ and $n \to \infty$. By the resolvent equation

$$U^{\alpha+\beta} g_n = U^\alpha g_n - \beta U^\alpha U^{\alpha+\beta} g_n = U^\alpha [g_n - \beta U^{\alpha+\beta} g_n] ,$$

and $g_n - \beta U^{\alpha+\beta} g_n$ is bounded and positive. Thus if we define $f_n = n[g_n - nU^{\alpha+n} g_n]$, then each $f_n \in b\underline{\underline{E}}^+$ and using the resolvent equation once again, $U^\alpha f_n = nU^{\alpha+n} g_n \uparrow f$.

(2.10) PROPOSITION. Let $f \in b\underline{\underline{E}}$. Then for each x, $\alpha \to U^\alpha f(x)$ is infinitely differentiable on $(0,\infty)$ and if $D = \frac{d}{d\alpha}$

(i) $D^n U^\alpha f = (-1)^n\, n!\, (U^\alpha)^{n+1} f$

(ii) $D^n(\alpha U^\alpha f) = (-1)^{n+1}\, n!\, (U^\alpha)^n [I - \alpha U^\alpha] f$.

PROOF. If $\alpha, \beta > 0$, $\alpha \neq \beta$, it follows from the resolvent equation that

$$\frac{U^\beta f - U^\alpha f}{\beta - \alpha} = -U^\beta U^\alpha f .$$

Letting $\beta \to \alpha$ and using (2.2-e) we obtain $DU^\alpha f = -(U^\alpha)^2 f$. For the induction step first note that

$$(2.11)\qquad (U^\alpha)^n f - (U^\beta)^n f = (U^\alpha - U^\beta)\, [(U^\alpha)^{n-1} + (U^\alpha)^{n-2} U^\beta + \cdots + (U^\beta)^{n-1}] f ,$$

and consequently if $\beta > \alpha$

$$|(U^\alpha)^n f - (U^\beta)^n f| \le \frac{n}{\alpha^n} \|f\|\, (U^\alpha - U^\beta) 1 \le \frac{n}{\alpha^{n+2}} \|f\|\, (\beta - \alpha) .$$

In particular, it follows from this inequality that $\alpha \to (U^\alpha)^n f(x)$ is continuous on $[\alpha_0, \infty)$ uniformly in x for each $\alpha_0 > 0$. Now from (2.11) and the resolvent equation

$$\frac{(U^\alpha)^n f - (U^\beta)^n f}{\beta - \alpha} = -U^\alpha U^\beta [(U^\alpha)^{n-1} + \cdots + (U^\beta)^{n-1}] f \to -n (U^\alpha)^{n+1} f$$

as $\beta \to \alpha$. Therefore

$$(2.12)\qquad D[(U^\alpha)^n f] = -n(U^\alpha)^{n+1} f ,$$

and this establishes (i). For (ii) observe that

$$D(\alpha U^\alpha f) = U^\alpha f + \alpha D(U^\alpha f) = U^\alpha f + \alpha(-(U^\alpha)^2 f) = U^\alpha [I - \alpha U^\alpha] f ,$$

where I is the identity. This is (ii) when $n = 1$ and the general result follows by induction from (2.12) because

$$D[(U^\alpha)^{n-1} f - \alpha (U^\alpha)^n f] = -(n-1)(U^\alpha)^n f - (U^\alpha)^n f + n\alpha (U^\alpha)^{n+1} f$$

$$= -n(U^\alpha)^n [I - \alpha U^\alpha] f .$$

3. RAY RESOLVENTS AND SEMIGROUPS

Throughout this section E denotes a compact metric space and $\underline{\underline{E}}$ denotes the σ-algebra of Borel subsets of E. Let $\underset{\sim}{C} = \underset{\sim}{C}(E)$ denote the Banach space of real valued continuous functions on E with the supremum norm. Let $(U^\alpha)_{\alpha>0}$ be a sub-Markov resolvent on $(E, \underline{\underline{E}})$ and let $\underset{\sim}{S}^\alpha = \mathcal{S}^\alpha \cap \underset{\sim}{C}(E)$ be the convex cone of continuous α-supermedian functions. As usual we shall write $\underset{\sim}{S} = \underset{\sim}{S}^0$. In addition, let $\underset{\sim}{S}^\infty = \bigcup_\alpha \underset{\sim}{S}^\alpha$; then each $\underset{\sim}{S}^\alpha$, $0 \le \alpha \le \infty$, is a convex cone contained in $\underset{\sim}{C}(E)$ and closed unded "$\wedge$".

(3.1) DEFINITION. (U^α) *is a Ray resolvent provided*:

(i) $U^\alpha \underset{\sim}{C} \subset \underset{\sim}{C}$ *for each* $\alpha > 0$.

(ii) $\underset{\sim}{S}^\infty$ *separates the points of* E.

(3.2) REMARK. $\underset{\sim}{S}^\infty$ separates the points of E if and only if $\underset{\sim}{S}^\alpha$ does so for some, and hence all, $\alpha > 0$. To see this it suffices to show that if $\underset{\sim}{S}^\infty$ separates, then so does $\underset{\sim}{S}^\alpha$ for each fixed $\alpha > 0$. Therefore fix $\alpha > 0$. If, $x, y \in E$, $x \ne y$, then because $\underset{\sim}{S}^\infty$ separates E there exists $f \in \underset{\sim}{S}^\beta$ for some $\beta > 0$ such that $f(x) \ne f(y)$. If $\beta \le \alpha$, then $f \in \underset{\sim}{S}^\beta \subset \underset{\sim}{S}^\alpha$. If $\beta > \alpha$, let $g = U^\alpha f$, $h = f + (\beta - \alpha) U^\alpha f$. Then either g or h separates x and y. Clearly $g \in \mathcal{S}^\alpha \cap \underset{\sim}{C} = \underset{\sim}{S}^\alpha$. Thus it suffices to show that $h \in \underset{\sim}{S}^\alpha$, and since h is obviously continuous, that $h \in \mathcal{S}^\alpha$. To this end

$$\gamma U^{\alpha+\gamma} h = \gamma U^{\alpha+\gamma} f + \gamma(\beta-\alpha) U^{\alpha+\gamma} U^\alpha f = \gamma U^{\alpha+\gamma} f + (\beta-\alpha)\,[U^\alpha f - U^{\alpha+\gamma} f]$$

$$= (\gamma + \alpha - \beta)\, U^{\alpha+\gamma} f + (\beta - \alpha)\, U^\alpha f .$$

If $\gamma + \alpha - \beta \le 0$, then $\gamma U^{\alpha+\gamma} h \le (\beta - \alpha)\, U^\alpha f \le h$, while if $\gamma + \alpha - \beta > 0$, then $(\alpha+\gamma) = (\gamma + \alpha - \beta) + \beta$ and so $(\gamma + \alpha - \beta) U^{\alpha+\gamma} f \le f$ since $f \in \mathcal{S}^\beta$. Therefore $\gamma U^{\alpha+\gamma} h \le h$ for all $\gamma > 0$, proving that $h \in \mathcal{S}^\alpha$.

(3.3) REMARK. If $\alpha > 0$, the vector space $\underset{\sim}{S}^\alpha - \underset{\sim}{S}^\alpha$ is dense in $\underset{\sim}{C}$. Since $\underset{\sim}{S}^\alpha$ separates E, so does $\underset{\sim}{S}^\alpha - \underset{\sim}{S}^\alpha$. Also $\underset{\sim}{S}^\alpha - \underset{\sim}{S}^\alpha$ contains the constants. If $a, b, c \in \mathbb{R}$, then $(a \wedge b) - c = (a - c) \wedge (b - c)$. Thus if $f, g, h, k \in \underset{\sim}{S}^\alpha$,

$$(3.4) \qquad \begin{aligned}(f - g) \wedge (h - k) &= [(f + k) - (g + k)] \wedge [(h + g) - (g + k)] \\ &= (f + k) \wedge (h + g) - (g + k).\end{aligned}$$

This implies that $\underset{\sim}{S}^\alpha - \underset{\sim}{S}^\alpha$ is closed under "$\wedge$" since $\underset{\sim}{S}^\alpha$ is. But $\underset{\sim}{S}^\alpha - \underset{\sim}{S}^\alpha$ is a vector space and so it also is closed under "$\vee$". Thus the lattice form of the Stone-Weierstrass theorem implies that $\underset{\sim}{S}^\alpha - \underset{\sim}{S}^\alpha$ is dense in $\underset{\sim}{C}$.

Suppose $(U^\alpha)_{\alpha > 0}$ is a resolvent on $(E, \underline{\underline{E}})$ that satisfies (3.1-i) and has the property that $\alpha U^\alpha f \to f$ pointwise as $\alpha \to \infty$ for each $f \in \underset{\sim}{C}$. Clearly (3.1-ii) holds in this case. Also it is easy to check that $\alpha U^\alpha f \to f$ in norm in this case.

(3.5) DEFINITION. *A sub-Markov semigroup $(P_t)_{t \ge 0}$ on $(E, \underline{\underline{E}})$ is a family of sub-Markov kernels P_t, $t \ge 0$ on $(E, \underline{\underline{E}})$ such that $P_{t+s} = P_t P_s$ for all $t, s \ge 0$.*

It is *not* assumed in (3.5) that $P_0 = I$. Of course, $P_0^2 = P_0$. If each P_t is a Markov kernel, then (P_t) is called a Markov semigroup.

(3.6) THEOREM. *Let $(U^\alpha)_{\alpha > 0}$ be a Ray resolvent. Then there exists a unique sub-Markov semigroup $(P_t)_{t \ge 0}$ satisfying:*

(i) *$t \to P_t f(x)$ is right continuous on $[0, \infty)$ for each $x \in E$ and $f \in \underset{\sim}{C}$.*

(ii) *$U^\alpha f = \int_0^\infty e^{-\alpha t} P_t f \, dt$ for $\alpha > 0$, $f \in \underset{\sim}{C}$.*

In addition

(iii) *$f \in \underset{\sim}{S}^\alpha$ if and only if $f \in \underset{\sim}{C}$ and $e^{-\alpha t} P_t f \le f$ for all $t \ge 0$.*

Moreover $e^{-\alpha t} P_t f \uparrow \hat{f} = P_0 f$ *as* $t \downarrow 0$.

(iv) *Let D be the set of $x \in E$ such that for each $f \in \underset{\sim}{C}$, $\alpha U^\alpha f(x) \to f(x)$ as $\alpha \to \infty$. Then D is Borel; $P_0(x, \cdot) = \epsilon_x$ if and only if $x \in D$ where ϵ_x is unit mass at x; and $P_t(x, \cdot)$ is carried by D for all $t \ge 0$ and $x \in E$.*

(v) *(P_t) is Markov if and only if (U^α) is Markov.*

PROOF. The uniqueness of (P_t) satisfying (i) and (ii) is immediate from the uniqueness theorem for Laplace transforms. We shall first prove the existence of (P_t) under the additional assumptions that $U = U^0$ is a bounded kernel with $U^0 \underset{\sim}{C} \subset \underset{\sim}{C}$ and such that $\underset{\sim}{S} = \underset{\sim}{S}^0 = \mathbf{\mathcal{S}} \cap \underset{\sim}{C}$ separates the points of E. This will represent the main work — the removal of these assumptions is easy. Under these assumptions it follows as in (3.3) that $\underset{\sim}{S} - \underset{\sim}{S}$ is (uniformly) dense in $\underset{\sim}{C}$.

If $f \in \underset{\sim}{S}$, then by (2.10)

$$(3.7) \qquad D^n[\alpha U^\alpha f] = (-1)^{n+1}\, n!\, (U^\alpha)^n\, [I - \alpha U^\alpha] f$$

where $D = \frac{d}{d\alpha}$. Recall that $\hat{f} = \lim_{\alpha\to\infty} \alpha U^\alpha f$ is the excessive regularization of f and define $g(\alpha) = \hat{f}(x) - \alpha U^\alpha f(x) \geq 0$. Then from (3.7),

$$(-1)^n D^n g(\alpha) = n!\, (U^\alpha)^n\, [I - \alpha U^\alpha] f(x) \geq 0 .$$

Thus g is completely monotone and $g(0) = g(0+) = \hat{f}(x)$ because $Uf(x) < \infty$. Consequently by the Hausdorff-Bernstein-Widder theorem (see [4], p. 439) there exists a positive measure $\lambda_x(f, \cdot)$ on $[0, \infty)$ of total mass $\hat{f}(x)$ such that

$$(3.8) \qquad \hat{f}(x) - \alpha U^\alpha f(x) = \int_{[0,\infty)} e^{-\alpha t}\, \lambda_x(f, dt)$$

for all $\alpha \geq 0$. But

$$\lambda_x(f, \{0\}) = \lim_{\alpha\to\infty} \int_{[0,\infty)} e^{-\alpha t}\, \lambda_x(f, dt) = \lim_{\alpha\to\infty} [\hat{f}(x) - \alpha U^\alpha f(x)] = 0 ,$$

by the very definition of $\hat{f}$, and so $\lambda_x(f, \cdot)$ is carried by $(0, \infty)$. Since $f \to \hat{f}(x) - \alpha U^\alpha f(x)$ is a cone map on $\underset{\sim}{S}$ (that is, it commutes with the taking of linear combinations with positive coefficients), so is $f \to \lambda_x(f, \cdot)$ by the uniqueness theorem for Laplace transforms. Now define

$$(3.9) \qquad P_t f(x) = \lambda_x(f, (t, \infty)) \quad \text{for} \quad f \in \underset{\sim}{S} .$$

Then for each $x \in E$, $t \to P_t f(x)$ is decreasing, right continuous, and $P_t f(x) \uparrow \hat{f}(x) = P_0 f(x)$ as $t \downarrow 0$ since $\lambda_x(f, \cdot)$ does not charge $\{0\}$. Moreover

$$\int_0^\infty e^{-\alpha t} P_t f(x)\,dt = \int_0^\infty e^{-\alpha t} \int_{(t,\infty)} \lambda_x(f, ds)\,dt$$

$$= \int_0^\infty \lambda_x(f, ds) \int_0^s e^{-\alpha t}\,dt = \frac{1}{\alpha}\int_0^\infty (1 - e^{-\alpha s})\,\lambda_x(f, ds)$$

$$= \frac{1}{\alpha}\,(\hat{f}(x) - [\hat{f}(x) - \alpha U^\alpha f(x)]) = U^\alpha f(x)\,.$$

Next for each $x \in E$ and $t \geq 0$ we extend $f \to P_t f(x)$ from $\underset{\sim}{S}$ to $\underset{\sim}{S} - \underset{\sim}{S}$ by linearity. This extension is again denoted by $P_t f(x)$. It is well defined since if $f - g = h - k$ with $f, g, h, k \in \underset{\sim}{S}$, then $f + k = h + g$ and

$$P_t f(x) + P_t k(x) = P_t(f + k)(x) = P_t(h + g)(x) = P_t h(x) + P_t g(x)\,.$$

It is easy to check that $f \to P_t f(x)$ is linear on $\underset{\sim}{S} - \underset{\sim}{S}$. Moreover by linearity $t \to P_t f(x)$ is right continuous and

$$\text{(3.10)} \qquad \int_0^\infty e^{-\alpha t} P_t f(x)\,dt = U^\alpha f(x)$$

for $f \in \underset{\sim}{S} - \underset{\sim}{S}$. If $f \in \underset{\sim}{S}$, then

$$D^n\left[\frac{f}{\alpha} - U^\alpha f\right] = \frac{(-1)^n n!}{\alpha^{n+1}}\, f - n!\,(-1)^n\,(U^\alpha)^{n+1} f = \frac{(-1)^n n!}{\alpha^{n+1}}\,[f - (\alpha U^\alpha)^{n+1} f]\,.$$

But iterating $\alpha U^\alpha f \leq f$, it follows that $\alpha \to \alpha^{-1} f - U^\alpha f$ is completely monotone, and clearly it is the Laplace transform of $f - P_t f$. Hence because of the right continuity of $t \to P_t f$ we see that $P_t f \leq f$ for all $f \in \underset{\sim}{S}$ and all $t \geq 0$. In particular, $P_t 1 \leq 1$. Next suppose $f \in \underset{\sim}{S} - \underset{\sim}{S}$ and $f \geq 0$. Then $D^n(U^\alpha f) = (-1)^n\, n!\,(U^\alpha)^{n+1} f$, and since $(U^\alpha)^{n+1} f \geq 0$ we see that $\alpha \to U^\alpha f$ is completely monotone. Hence from (3.10) and the right continuity of $t \to P_t f$ we obtain $P_t f \geq 0$ for all $t \geq 0$.

Therefore $f \to P_t f(x)$ is a positive, bounded (by 1), linear form on $\underset{\sim}{S} - \underset{\sim}{S}$, and so may be extended by continuity to a positive, bounded (by 1), linear form on $\underset{\sim}{C}$ which is the uniform closure of $\underset{\sim}{S} - \underset{\sim}{S}$. If we denote this extension once again by $P_t f(x)$, then by the Riesz representation theorem there exists a subprobability measure, $P_t(x, \cdot\,)$ on $(E, \underline{\underline{E}})$ such that

(3.11) $$P_t f(x) = \int P_t(x, dy)\, f(y)$$

for all $t \geq 0$, $x \in E$, and $f \in \underset{\sim}{C}$. Moreover if $(f_n) \subset \underset{\sim}{S} - \underset{\sim}{S}$ and $f_n \to f$ uniformly, then $\| P_t f - P_t f_n \| \leq \| f - f_n \|$ and $\| U^\alpha f - U^\alpha f_n \| \leq \alpha^{-1} \| f - f_n \|$. As a result of this and the fact that $\underset{\sim}{S} - \underset{\sim}{S}$ is dense in $\underset{\sim}{C}$, (3.10) holds for all $f \in \underset{\sim}{C}$, $x \in E$, and $\alpha > 0$. Moreover the right continuity of $t \to P_t f(x)$ for $f \in \underset{\sim}{S} - \underset{\sim}{S}$ implies that it is also right continuous for all $f \in \underset{\sim}{C}$. Letting $\alpha \downarrow 0$ and using the boundedness of $U = U^0$, (3.10) holds even when $\alpha = 0$.

We interrupt the main argument to state and prove a lemma that will be used several times in the sequel.

(3.12) LEMMA. *Let* $(E, \underline{\underline{E}})$ *be a measurable space and let* $g: \mathbb{R}^+ \times E \to \mathbb{R}$. *Suppose that* $t \to g(t,x)$ *is right continuous on* $\mathbb{R}^+$ *for each* x *in* E *and that there exist a positive* β *and a positive Borel function* g_0 *on* $\mathbb{R}^+$ *such that* $|g(t,x)| \leq g_0(t)$ *and* $\int_0^\infty e^{-\beta t} g_0(t)\, dt < \infty$. *Let* $h(\alpha, x) = \int_0^\infty e^{-\alpha t} g(t,x)\, dt$ *for* $\alpha \geq \beta$. *If* $x \to h(\alpha, x)$ *is* $\underline{\underline{E}}$ *measurable for each* $\alpha > \beta$, *then* $x \to g(t,x)$ *is* $\underline{\underline{E}}$ *measurable for each* $t \geq 0$ *and* $(t,x) \to g(t,x)$ *is* $\underline{\underline{R}}^+ \otimes \underline{\underline{E}}$ *measurable.*

PROOF. It follows from the hypothesis that if ϕ is a polynomial in e^{-t}, then $x \to \int_0^\infty e^{-\beta t} \phi(t)\, g(t,x)\, dt$ is $\underline{\underline{E}}$ measurable. Consequently the locally compact version of the Stone-Weierstrass theorem implies that this same statement is true whenever ϕ is a continuous function on $\mathbb{R}^+$ which vanishes at infinity. For $\tau \in \mathbb{R}^+$ given, define $\phi_n(t)$ to vanish outside the interval $(\tau, \tau + 1/n)$, $\phi_n(\tau + 1/2n) = 2n$, and $\phi_n(t)$ to be linear on each of the intervals $[\tau, \tau + 1/2n]$ and $[\tau + 1/2n, \tau + 1/n]$. Thus ϕ_n is continuous, $\phi_n \geq 0$, ϕ_n vanishes outside the interval $(\tau, \tau + 1/n)$, and $\int \phi_n(t)\, dt = 1$. Using the right continuity of $t \to g(t,x)$ it is easy to check that

$$\lim_{n \to \infty} \int_0^\infty e^{-\beta t} \phi_n(t)\, g(t,x)\, dt = e^{-\beta \tau} g(\tau, x),$$

for each x in E. But each integral is $\underline{\underline{E}}$ measurable and hence so is $x \to g(\tau, x)$. Since $\tau \in \mathbb{R}^+$ is arbitrary, this proves the first assertion (3.12); the second is a standard consequence of the first.

We return now to the proof of (3.6). If $f \in b\underline{\underline{E}}$ we define $P_t f(x) =$

$\int P_t(x, dy)\, f(y)$. According to (3.11) this agrees with our previous definition when $f \in \underset{\sim}{C}$. It is immediate from (3.10) and (3.12) that $(t, x) \to P_t f(x)$ is $\underline{\underline{R}}^+ \otimes \underline{\underline{E}}$ measurable for each $f \in \underset{\sim}{C}$, and a standard monotone class argument then implies that the statement is true for each $f \in b\underline{\underline{E}}$. In particular for each $t \geq 0$, P_t is a kernel on $(E, \underline{\underline{E}})$. The left side of (3.10) now makes sense for $f \in b\underline{\underline{E}}$, and since both sides are measures in f, (3.10) holds for all $f \in b\underline{\underline{E}}$.

To complete the proof of existence under the present hypotheses it remains to show that $(P_t)_{t \geq 0}$ forms a semigroup. Since $P_t P_s$ and P_{t+s} are subprobability measures it suffices to show $P_t P_s f = P_{t+s} f$ for all $f \in \underset{\sim}{C}^+$ and $t, s \geq 0$. To this end fix $f \in \underset{\sim}{C}^+$. Then both $P_t P_s f$ and $P_{t+s} f$ are right continuous in s, and so it suffices to show that for each fixed $t \geq 0$ they have the same Laplace transform on s. That is, after an obvious change of order of integration,

$$(3.13) \qquad P_t U^\alpha f(x) = \int_0^\infty e^{-\alpha s} P_{t+s} f(x)\, ds \; .$$

But for fixed $\alpha > 0$ the right side of (3.13) is right continuous in t, and since $U^\alpha f \in \underset{\sim}{C}$ so is the left side. Thus (3.13) will hold identically in t provided both sides have the same Laplace transform on t. Thus the proof of the semigroup property reduces to verifying

$$(3.14) \qquad U^\beta U^\alpha f(x) = \int_0^\infty e^{-\beta t} \int_0^\infty e^{-\alpha s} P_{t+s} f(x)\, ds\, dt \; .$$

The right side of (3.14) is equal to

$$\int_0^\infty e^{-\beta t} e^{\alpha t} \int_t^\infty e^{-\alpha s} P_s f\, ds\, dt = \int_0^\infty e^{-\alpha s} P_s f \int_0^s e^{-(\beta - \alpha)t}\, dt\, ds$$

$$= (\beta - \alpha)^{-1} \int_0^\infty (e^{-\alpha s} - e^{-\beta s}) P_s f\, ds = (\beta - \alpha)^{-1} [U^\alpha - U^\beta] f = U^\beta U^\alpha f$$

provided $\beta \neq \alpha$. This actually is enough to conclude from (3.14) that (P_t) forms a semigroup. However, if $\beta = \alpha$ the same computation shows that the right side of (3.14) equals

$$\int_0^\infty s e^{-\alpha s} P_s f \, ds = -D \int_0^\infty e^{-\alpha s} P_s f \, ds = -DU^\alpha f = (U^\alpha)^2 f ,$$

which establishes (3.14) when $\beta = \alpha$. This completes the proof of the existence of the semigroup $(P_t)_{t \geq 0}$ satisfying (3.6-i) and (3.6-ii) under the special assumptions stated in the second sentence of the proof.

For the general case for each $\beta > 0$ define $V_\beta^\alpha = U^{\alpha+\beta}$. Then $(V_\beta^\alpha)_{\alpha>0}$ is a Ray resolvent with $V_\beta^0 = U^\beta$ bounded and, using an obvious notation, $\underset{\sim}{S}^0(V_\beta) = \underset{\sim}{S}^\beta(U)$ separates the points of E. Thus by what has already been proved there exists for each $\beta > 0$ a semigroup (P_t^β) of sub-Markov kernels on $(E, \underline{\underline{E}})$ such that $t \to P_t^\beta f(x)$ is right continuous for $f \in \underset{\sim}{C}$ and satisfying

$$(3.15) \qquad U^{\alpha+\beta} f(x) = \int_0^\infty e^{-\alpha t} P_t^\beta f(x) \, dt$$

for $\alpha > 0$. Let

$$g(\alpha) = \int_0^\infty e^{-\alpha t} [e^{-\beta t} - P_t^\beta 1(x)] \, dt = \frac{1}{\alpha+\beta} - U^{\alpha+\beta} 1(x) .$$

As before it is easy to check that $(-1)^n D^n g \geq 0$, and again this together with the right continuity of $t \to P_t^\beta 1$ imply that $P_t^\beta 1 \leq e^{-\beta t}$ for all t. Now define $P_t = e^{\beta t} P_t^\beta$ where $\beta > 0$ is fixed. It is clear that $(P_t)_{t \geq 0}$ is a sub-Markov semigroup, and for $f \in \underset{\sim}{C}$, (3.15) gives

$$(3.16) \qquad \int_0^\infty e^{-\alpha t} P_t f \, dt = \int_0^\infty e^{-\alpha t} e^{\beta t} P_t^\beta f \, dt = U^\alpha f$$

if $\alpha > \beta$. Thus by the uniqueness theorem for Laplace transforms $P_t f$ does not depend on β, and since $\beta > 0$ is arbitrary in (3.16) we see that $(P_t)_{t \geq 0}$ satisfies (3.6-i) and (3.6-ii).

To check (iii) observe that if $f \in \underset{\sim}{C}$, then $g(\beta) = \beta^{-1} f - U^{\alpha+\beta} f$ is the Laplace transform of $f - e^{-\alpha t} P_t f$. Thus if $e^{-\alpha t} P_t f \leq f$ for all t, then $f \in \underset{\sim}{S}^\alpha$. Conversely, if $f \in \underset{\sim}{S}^\alpha$ one checks as usual that g is completely monotone which implies $e^{-\alpha t} P_t f \leq f$ for all t. It is a standard (and easy) fact about Laplace transforms that

$$P_0 f = \lim_{t \downarrow 0} e^{-\alpha t} P_t f = \lim_{\beta \to \infty} \beta U^{\alpha+\beta} f = \hat{f}$$

for $f \in S^{\alpha}$. This proves (iii). For (v), apply (iii) with $f = 1$ and $\alpha = 0$ to see that $\beta^{-1} - U^{\beta} 1$ is the Laplace transform of $1 - P_t 1$ which yields (v).

Finally, we turn to (iv). Since $\alpha U^{\alpha} = \alpha U^{\alpha+\beta} + \alpha \beta U^{\alpha} U^{\alpha+\beta}$, it is clear that for $f \in \underset{\sim}{C}$, $\alpha U^{\alpha} f(x) \to f(x)$ as $\alpha \to \infty$ if and only if $\alpha U^{\alpha+\beta} f(x) \to f(x)$ as $\alpha \to \infty$ for some, and hence all, $\beta \geq 0$. In particular $\alpha U^{\alpha+1} f(x) \to f(x)$ for $x \in D$ and $f \in \underset{\sim}{C}$. Consequently if $x \in D$, then $f(x) = \hat{f}(x)$ for all $f \in \underset{\sim}{S}^1$ where, as usual, $\hat{f}$ is the 1-excessive regularization of f. Let $\{g_n\}$ be a countable dense subset of $\underset{\sim}{S}^1$ in the uniform norm. (This is possible since $\underset{\sim}{S}^1$ is a subset of the separable Banach space $\underset{\sim}{C}$.) Fix $x \in E$ and suppose that $g_n(x) = \hat{g}_n(x)$ for all n. Since $\{g_n\}$ is dense in $\underset{\sim}{S}^1$ it follows that $g(x) = \hat{g}(x)$ for all $g \in \underset{\sim}{S}^1$. Next let $\mathcal{K} = \{f \in \underset{\sim}{C}: \alpha U^{\alpha} f(x) \to f(x) \text{ as } \alpha \to \infty\}$. Then it is evident that $\mathcal{K}$ is a closed linear subspace of $\underset{\sim}{C}$ that contains $\underset{\sim}{S}^1$ and hence $\underset{\sim}{S}^1 - \underset{\sim}{S}^1$. But $\underset{\sim}{S}^1 - \underset{\sim}{S}^1$ is dense in $\underset{\sim}{C}$ and so $\mathcal{K} = \underset{\sim}{C}$. That is $x \in D$. Thus we have shown that

(3.17) $$D = \bigcap_n \{x: g_n(x) = \hat{g}_n(x)\} .$$

Consequently D is Borel. If $g \in \underset{\sim}{S}^1$, then $P_0 g = \lim_{t \to 0} e^{-t} P_t g = \lim_{\alpha \to \infty} \alpha U^{\alpha+1} g = \hat{g}$. Thus if $P_0(x, \cdot) = \varepsilon_x$, then $x \in D$. Conversely if $x \in D$, then $P_0 g(x) = \hat{g}(x) = g(x)$ for all $g \in \underset{\sim}{S}^1$, and, as before, this easily implies that $P_0(x, \cdot) = \varepsilon_x$. If $g \in \underset{\sim}{S}^1$, then $P_t g = P_t P_0 g = P_t \hat{g}$, and so $P_t(g - \hat{g}) = 0$. Since $\hat{g} \leq g$, $P_t(x, \cdot)$ is carried by $\{g = \hat{g}\}$, and in view of (3.17) this states that $P_t(x, \cdot)$ is carried by D for all $t \geq 0$ and $x \in E$. This completes the proof of Theorem 3.6.

(3.18) REMARKS. (i) It is immediate from (3.6-ii) — or more accurately this relationship for $f \in b\underline{\underline{E}}$ — and (3.6-iv) that $U^{\alpha}(x, \cdot)$ is carried by D for each $\alpha > 0$ and $x \in E$.

(ii) The Borel set $B = E - D$ is called the set of branch points of $(U^{\alpha})_{\alpha > 0}$, or of $(P_t)_{t \geq 0}$.

(iii) If B is empty, then for each $f \in \underset{\sim}{C}$, $\alpha U^{\alpha} f \to f$ pointwise. As mentioned below (3.4), $\| \alpha U^{\alpha} f - f \| \to 0$ in this case. Moreover it can be shown in this situation that each P_t maps $\underset{\sim}{C}$ into $\underset{\sim}{C}$. The reader may find it instructive to give an example of a Ray resolvent (U^{α}) for which the corresponding semigroup (P_t) does not send $\underset{\sim}{C}$ into $\underset{\sim}{C}$.

In discussing Ray resolvents we shall often consider U^{α} and P_t as operators on $b\underline{\underline{E}}^*$ or $\underline{\underline{E}}^*_+$. Both U^{α} and P_t map $b\underline{\underline{E}}^*$ into itself, and the same statement holds for $\underline{\underline{E}}^*_+$. If λ and μ are finite measures on $(\mathbb{R}^+, \underline{\underline{R}}^+)$ and $(E, \underline{\underline{E}})$ respectively let $(\underline{\underline{R}}^+ \otimes \underline{\underline{E}})^{\lambda,\mu}$ denote the completion of the product σ-algebra $\underline{\underline{R}}^+ \otimes \underline{\underline{E}}$ with respect to the product measure $\lambda \otimes \mu$. If $f \in b\underline{\underline{E}}^*$ it is easy to see that $(t,x) \to P_t f(x)$ is $(\underline{\underline{R}}^+ \otimes \underline{\underline{E}})^{\lambda,\mu}$ measurable for each λ and μ. Now if $f \in b\underline{\underline{E}}$

(3.19) $$U^{\alpha} f(x) = \int e^{-\alpha t} P_t f(x)\, dt ,$$

and since both sides are measures in f it follows that (3.19) holds for all $f \in b\underline{\underline{E}}^*$.

There is a standard device for reducing the sub-Markov case to the Markov case which we shall now explain. See also BG-page 46. Suppose $(U^{\alpha})_{\alpha>0}$ is a sub-Markov resolvent on $(E, \underline{\underline{E}})$. Let Δ be a point not in E and define $E_{\Delta} = E \cup \{\Delta\}$. Let Δ be isolated in E_{Δ} so E_{Δ} is a compact metric space. Define

$$V^{\alpha}(x, A) = U^{\alpha}(x, A);\; x \in E,\; A \in \underline{\underline{E}}$$

$$V^{\alpha}(x, \{\Delta\}) = \alpha^{-1} - U^{\alpha}(x, E);\; x \in E$$

$$V^{\alpha}(\Delta, \cdot) = \alpha^{-1} \varepsilon_{\Delta}.$$

One quickly checks that $(V^{\alpha})_{\alpha>0}$ is a Markov resolvent on $(E_{\Delta}, \underline{\underline{E}}_{\Delta})$, and (V^{α}) is a Ray resolvent on E_{Δ} if and only if (U^{α}) is a Ray resolvent on E. If $(Q_t)_{t\geq 0}$ is the semigroup constructed from V and $(P_t)_{t\geq 0}$ the semigroup constructed from U, then for all $t \geq 0$

$$Q_t(x, A) = P_t(x, A);\; x \in E,\; A \in \underline{\underline{E}}$$

$$Q_t(x, \{\Delta\}) = 1 - P_t(x, E);\; x \in E$$

$$Q_t(\Delta, \cdot) = \varepsilon_{\Delta}.$$

4. INCREASING SEQUENCES OF SUPERMARTINGALES

Before proceeding to the main business at hand, we shall prove in this section a very useful result about increasing sequences of supermartingales that is due to Meyer. See [8], Ch. VI, T16. However, we shall give a proof using D-IV-T28. In the statement of the following theorem $(\Omega, \underline{\underline{F}}, \underline{\underline{F}}_t, P)$ satisfies the usual hypotheses of the general theory, D-III-26, and, of course, all supermartingales are relative to the family $(\underline{\underline{F}}_t)$.

(4.1) THEOREM. Let (X^n) be a sequence of supermartingales satisfying:

(i) For each n, $t \to X_t^n$ is right continuous almost surely.

(ii) For each n and t, $P[X_t^n > X_t^{n+1}] = 0$.

Let $X_t(\omega) = \sup_n X_t^n(\omega)$. Then almost surely $t \to X_t(\omega)$ is right continuous and has left limits in $(-\infty, \infty]$.

PROOF. It follows from (i) and (ii) that there exists a set $\Lambda \in \underline{\underline{F}}$ with $P(\Lambda) = 0$ such that for $\omega \notin \Lambda$, $t \to X_t^n(\omega)$ is right continuous for each n and $X_t^n(\omega) \leq X_t^{n+1}(\omega)$ for each t and n. By redefining $X_t^n(\omega) = 0$ for $\omega \in \Lambda$ we may assume without loss of generality that for each ω and n, $t \to X_t^n(\omega)$ is right continuous and $X_t^n(\omega) \leq X_t^{n+1}(\omega)$ for all n, t, and ω. Of course, $X_t(\omega) = \sup_n X_t^n(\omega)$ may be infinite and so the conclusion of the theorem must be interpreted in the topology of $(-\infty, \infty]$. After these preliminary remarks we turn to the proof of (4.1).

Suppose firstly that there is a constant K such that $0 \leq X_t^n(\omega) \leq K$ for all t, n, and ω. Then $0 \leq X_t(\omega) \leq K$ and $X = (X_t)$ is well measurable since each X^n is. Moreover, if $0 \leq s < t$, then letting $n \to \infty$ in $X_s^n \geq E(X_t^n \mid \underline{\underline{F}}_s)$ we see that X is a supermartingale. Since $X_t^n(\omega) \uparrow X_t(\omega)$ for all n, t, and ω, it follows that if S is a bounded stopping time, $X_S^n \uparrow X_S$. Now let (S_k) be a decreasing sequence of bounded stopping times with limit S. Then $E(X_{S_k}^n)$ increases with both n and k, and so using the right continuity of X^n for each n

we have

$$\lim_k E(X_{S_k}) = \lim_k \lim_n E(X^n_{S_k}) = \lim_n \lim_k E(X^n_{S_k})$$

$$= \lim_n E(X^n_S) = E(X_S) \ .$$

Consequently by D-IV-T28, X is almost surely right continuous, and since it is also a supermartingale it must have left limits almost surely.

For the general case fix $a > 0$. Let $M = (M_t)$ be a right continuous version of the martingale $E(X^1_a | \underline{\underline{F}}_t)$. Then almost surely M has left limits and almost surely $M_t \le X^1_{t\wedge a}$ with equality if $t \ge a$. Let $q(x) = x(1+x)^{-1}$ if $0 \le x < \infty$ and $q(\infty) = 1$. Then q is an order preserving homeomorphism of $[0, \infty]$ on $[0, 1]$. Define $Y^n_t = q(X^n_{t\wedge a} - M_t)$. By Jensen's inequality Y^n is a right continuous supermartingale satisfying $0 \le Y^n_t(\omega) \le 1$. By the first part of the proof $Y_t = \sup_n Y^n_t$ is right continuous and has left limits almost surely. But $Y_t = q(X_{t\wedge a} - M_t)$ and so $X_t - M_t$, and hence X_t, is right continuous and has left limits on $[0, a]$ almost surely. Letting $a \to \infty$ through a sequence completes the proof.

(4.2) REMARK. In later sections we shall apply (4.1) when the system $(\Omega, \underline{\underline{F}}, \underline{\underline{F}}_t, P)$ does not satisfy the usual hypotheses. To be precise let $(\Omega, \underline{\underline{F}}^0, \underline{\underline{F}}^0_t, P)$ be a system in which it is only assumed that $(\underline{\underline{F}}^0_t)$ is increasing. Let $\underline{\underline{F}}$ be the completion of $\underline{\underline{F}}^0$ with respect to P and let $\underline{\underline{F}}_t$ be the σ-algebra generated by $\underline{\underline{F}}^0_{t+}$ and all sets in $\underline{\underline{F}}$ of P measure zero. It is easy to check that $(\Omega, \underline{\underline{F}}, \underline{\underline{F}}_t, P)$ satisfies the usual conditions. Moreover if $X^n = (X^n_t)$ is almost surely right continuous and a supermartingale over $(\Omega, \underline{\underline{F}}^0, \underline{\underline{F}}^0_t, P)$, then using the right continuity of X^n one checks that X^n is a supermartingale over $(\Omega, \underline{\underline{F}}, \underline{\underline{F}}_t, P)$. See, for example, VI-T4 of [8]. In this manner one reduces the statement of (4.1) relative to the system $(\Omega, \underline{\underline{F}}^0, \underline{\underline{F}}^0_t, P)$ to the statement relative to $(\Omega, \underline{\underline{F}}, \underline{\underline{F}}_t, P)$. Thus in the sequel we shall use (4.1) whether or not the underlying system satisfies the usual hypotheses.

5. PROCESSES

Throughout this section $(U^\alpha)_{\alpha>0}$ will be a fixed Ray resolvent on a compact metric space E and $(P_t)_{t\geq 0}$ will be the semigroup constructed in Section 3. As explained at the end of Section 3 there is no real loss of generality in assuming that (U^α) and (P_t) are Markov. Thus we shall assume that $\alpha U^\alpha 1 = 1$ for each $\alpha > 0$ and that $P_t 1 = 1$ for each $t \geq 0$. As in Section 3, B denotes the (Borel) set of branch points and $D = E - B$. Of course, $\underline{\underline{E}}$ is the σ-algebra of Borel subsets of E. Recall that $\mathbb{R}^+ = [0, \infty)$ and $\mathbb{R}^{++} = (0, \infty)$.

Let W be the set of all right continuous functions $w: \mathbb{R}^+ \to D$ that have left limits in E at each point of $\mathbb{R}^{++}$. Let $Y_t(w) = w(t)$. We emphasize that for each $t \geq 0$, $Y_t(w) \in D$, and that for each $t > 0$, $Y_{t-}(w)$ exists as a point in E. Let $\underline{\underline{G}}^0 = \sigma(Y_t;\, t \geq 0)$ and $\underline{\underline{G}}_t^0 = \sigma(Y_s;\, s \leq t)$. (If $\mathcal{H}$ is a collection of functions from a set Ω to a measurable space $(E, \underline{\underline{E}})$, then $\sigma(\mathcal{H})$ denotes the smallest σ-algebra of subsets of Ω relative to which each element of $\mathcal{H}$ is measurable; that is, the σ-algebra generated by $\mathcal{H}$.)

(5.1) THEOREM. *For each probability* μ *on* $(E, \underline{\underline{E}})$, *there exists a probability* P^μ *on* $(W, \underline{\underline{G}}^0)$ *such that* $(Y_t, \underline{\underline{G}}_t^0, P^\mu)$ *is a Markov process with transition semigroup* (P_t) *and initial measure* μP_0; *that is*

(i) $E^\mu[f(Y_{t+s}) \mid \underline{\underline{G}}_s^0] = P_t f(Y_s)$, *for* $f \in b\underline{\underline{E}}$; $t, s \geq 0$;

(ii) $P^\mu[Y_0 \in A] = \int \mu(dx)\, P_0(x, A)$, *for* $A \in \underline{\underline{E}}$.

PROOF. We shall only sketch the familiar argument. See BG-pages 47-49. Let $\Omega = E^{\mathbb{R}^+}$ and $\underline{\underline{F}} = \underline{\underline{E}}^{\mathbb{R}^+}$ so that $(\Omega, \underline{\underline{F}})$ is the usual product measurable space. Let $X_t(\omega) = \omega(t)$, $t \in \mathbb{R}^+$ be the coordinate maps and let $\underline{\underline{F}}_t = \sigma(X_s;\, s \leq t)$. Then given μ one can apply the Kolmogorov extension theorem to obtain a measure P on $(\Omega, \underline{\underline{F}})$ such that $(X_t, \underline{\underline{F}}_t, P)$ is a Markov process with transition semigroup (P_t) and initial measure μP_0. In particular for $f \in b\underline{\underline{E}}$ and

$t \geq 0$, $E[f(X_t)] = \int \mu(dx)\, P_t f(x)$.

We shall show that this process has a modification that is right continuous and has left limits. Recall from Section 3 that $\underset{\sim}{S}^{\alpha}$ is the collection of continuous α-supermedian functions. By (3.6-iii) if $f \in \underset{\sim}{S}^{\alpha}$, $e^{-\alpha t} P_t f \leq f$. Consequently

$$E[e^{-\alpha(t+s)} f(X_{t+s}) \mid \underset{=}{F}_s] = e^{-\alpha s}\, e^{-\alpha t}\, P_t f(X_s) \leq e^{-\alpha s} f(X_s) .$$

Therefore $(e^{-\alpha t} f(X_t), \underset{=}{F}_t, P)$ is a positive supermartingale and so almost surely P, $t \to f(X_t)$ restricted to the positive rationals, $\underset{\sim}{Q}^+$, has left and right limits at each point in $\mathbb{R}^+$. (See, for example, VI-T3 of [8].) Fix $\alpha > 0$, say $\alpha = 1$ for convenience. Since $\underset{\sim}{S}^1 - \underset{\sim}{S}^1$ is dense in $\underset{\sim}{C} = \underset{\sim}{C}(E)$, it follows that for each $f \in \underset{\sim}{C}$ almost surely P, $t \to f(X_t)$ restricted to $\underset{\sim}{Q}^+$ has right and left limits at each point of $\mathbb{R}^+$. Since $\underset{\sim}{C}$ is separable this implies that almost surely P, $t \to X_t$ restricted to $\underset{\sim}{Q}^+$ has right and left limits in E at each point of $\mathbb{R}^+$. By throwing out of Ω a set of P measure zero one may assume that for each $\omega \in \Omega$, $t \to X_t(\omega)$ restricted to $\underset{\sim}{Q}^+$ has right and left limits at each point of $\mathbb{R}^+$. For each $t \in \mathbb{R}^+$ and $\omega \in \Omega$ define

$$X_{t+}(\omega) = \lim_{r \downarrow t,\, r \in \underset{\sim}{Q}^+} X_r(\omega) .$$

If $f, g \in \underset{\sim}{S}^1$, $t \in \mathbb{R}^+$, and $(r_n) \subset \underset{\sim}{Q}^+$ with $r_n \downarrow t$, $r_n > t$, then using the Markov property

$$(5.2) \quad E[f(X_t)\, g(X_{t+})] = \lim_n E[f(X_t)\, g(X_{r_n})]$$

$$= \lim_n E[f(X_t)\, P_{r_n - t}\, g(X_t)] = E[f(X_t)\, \hat{g}(X_t)]$$

since $P_s g \to P_0 g = \hat{g}$ as $s \downarrow 0$. Recall that $\hat{g}$ is the 1-excessive regularization of g. But $P_t g = P_t P_0 g = P_t \hat{g}$, and so

$$E[g(X_t) - \hat{g}(X_t)] = \int \mu(dx)\, P_t(g - \hat{g})(x) = 0 .$$

Since $0 \leq \hat{g} \leq g$ this implies that $g(X_t) = \hat{g}(X_t)$ almost surely, and combining this with (5.2) we obtain

$$(5.3) \quad E[f(X_t)\, g(X_{t+})] = E[f(X_t)\, g(X_t)]$$

for $f, g \in \underset{\sim}{S}^1$. Since $\underset{\sim}{S}^1 - \underset{\sim}{S}^1$ is dense in $\underset{\sim}{C}$, (5.3) must hold for all $f, g \in \underset{\sim}{C}$, and then a standard argument shows that

$$\text{(5.4)} \qquad E[F(X_t, X_{t+})] = E[F(X_t, X_t)]$$

for all $F \in b(\underline{\underline{E}} \otimes \underline{\underline{E}})$. Taking $F(x,y)$ to be the indicator of the diagonal in $E \times E$ yields $P[X_{t+} \neq X_t] = 0$ for each $t \geq 0$.

Next define $Z_t(\omega) = X_{t+}(\omega)$ for $t \in \mathbb{R}^+$ and $\omega \in \Omega$. Then $t \to Z_t(\omega)$ is right continuous on $\mathbb{R}^+$ and has left limits on $\mathbb{R}^{++}$. Since $Z_t = X_t$ almost surely for each $t \geq 0$, it follows that if $\underline{\underline{H}}_t = \sigma(Z_s;\, s \leq t)$, then $(Z_t, \underline{\underline{H}}_t, P)$ is a Markov process with transition semigroup (P_t) and initial measure μP_0. Of course, (Z_t) takes its values in E and its left limits exist in E. We now claim that there exists a P-null set Λ, such that if $\omega \notin \Lambda$, then $Z_t(\omega) \in D$ for all $t \geq 0$. To this end if $g \in \underset{\sim}{S}^1$, define $g_n = n[g - nU^{n+1} g]$. Then $g_n \in \underset{\sim}{C}^+$ and $f_n = U^1 g_n = nU^{n+1} g \uparrow \hat{g}$ as $n \to \infty$. Since $f_n \in \underset{\sim}{S}^1$, $\{e^{-t} f_n(Z_t), \underline{\underline{H}}_t, P)$ is a right continuous, positive supermartingale for each n, and hence by Theorem 4.1, $t \to \hat{g}(Z_t)$ is right continuous and has left limits almost surely. But we have already seen that for each fixed t, $g(Z_t) = \hat{g}(Z_t)$ almost surely, and because both $t \to g(Z_t)$ and $t \to \hat{g}(Z_t)$ are right continuous almost surely, it follows that these two processes are indistinguishable. On the other hand by (3.17), $D = \cap \{g_n = \hat{g}_n\}$ where $\{g_n\}$ is a countable dense subset of $\underset{\sim}{S}^1$, and so

$$P[\text{there exists } t \geq 0 \text{ with } Z_t \notin D] = 0 .$$

Thus deleting one more set of measure zero we may assume that $Z_t(\omega) \in D$ for all $t \geq 0$ and $\omega \in \Omega$. The process $(Z_t, \underline{\underline{H}}_t, P)$ now has all of the properties claimed in Theorem 5.1, and one just carries the measure P over to W where we call it P^μ. More precisely if $\pi : \Omega \to W$ is defined by $\pi(\omega)(t) = Z_t(\omega)$, then $P^\mu = \pi(P)$, and it is easy to check that P^μ has all of the desired properties. See, for example, BG-I-(4.3). This completes the proof of Theorem 5.1.

As is standard one introduces the shift operators $(\theta_t)_{t \geq 0}$ on W by $(\theta_t w)(s) = w(t+s)$, or equivalently $Y_t \circ \theta_s = Y_{t+s}$ for all $t, s \geq 0$. We write P^x for P^{ϵ_x}. For each probability μ on $(E, \underline{\underline{E}})$ let $\underline{\underline{G}}^\mu$ denote the completion of $\underline{\underline{G}}^0$ with respect to P^μ and let $\underline{\underline{N}}^\mu = \{\Lambda \in \underline{\underline{G}}^\mu : P^\mu(\Lambda) = 0\}$. Define $\underline{\underline{G}}_t^\mu = \sigma(\underline{\underline{G}}_t^0 \cup \underline{\underline{N}}^\mu)$ and $\underline{\underline{G}}_t = \cap\, \underline{\underline{G}}_t^\mu$, $\underline{\underline{G}} = \cap\, \underline{\underline{G}}^\mu$ where the intersections are over all probabilities μ on $(E, \underline{\underline{E}})$. The Markov property (5.1-i) extends as follows

$$E^{\mu}[Z \circ \theta_t \mid \underline{\underline{G}}^{\mu}_t] = E^{Y(t)}(Z) \tag{5.5}$$

for $Z \in b\underline{\underline{G}}$, $t \geq 0$, and all μ. The expression on the right side of (5.5) is the evaluation of the function $x \to E^x(Z)$ at the point Y_t. For typographical reasons we write $Y(t)$ for Y_t whenever convenient. Actually $t \to E^{Y(t)}(Z)$ is $\underline{\underline{G}}_t$ measurable. In particular, if $f \in b\underline{\underline{E}}^*$

$$E^{\mu}[f(Y_{t+s}) \mid \underline{\underline{G}}^{\mu}_s] = P_t f(Y_s) .$$

These facts are proved in BG-I-Sec. 5 to which we refer the reader.

Note that for each μ, the system $(W, \underline{\underline{G}}^{\mu}, \underline{\underline{G}}^{\mu}_t, P^{\mu})$ satisfies the usual hypotheses of the general theory of processes. (The right continuity of the family $(\underline{\underline{G}}^{\mu}_t)_{t\geq 0}$ is established in Theorem 5.8 below.) A statement such as $Z = (Z_t)$ is P^{μ} well measurable means that the process Z is well measurable relative to this system. P^{μ} previsible and P^{μ} indistinguishable have similar meanings. Two processes Z^1 and Z^2 will be called indistinguishable if they are P^{μ} indistinguishable for each μ. Similarly a property is said to hold almost surely provided that it holds P^{μ} almost surely for each μ.

We shall now change our terminology slightly and say that $f \geq 0$ is α-excessive (for Y) if $f \in \underline{\underline{E}}^*_+$ and $\beta U^{\alpha+\beta} f \uparrow f$ as $\beta \uparrow \infty$. Similarly we shall say that $f \in \underline{\underline{E}}^*_+$ is α-supermedian if $\beta U^{\alpha+\beta} f \leq f$ for all $\beta > 0$. In the terminology of Section 2 this amounts to saying that f is α-excessive (α-supermedian) for the resolvent $(U^{\alpha})_{\alpha > 0}$ considered as a resolvent on $(E, \underline{\underline{E}}^*)$ rather than on $(E, \underline{\underline{E}})$. We let $\mathcal{E}^{\alpha}$ (resp. $\mathcal{S}^{\alpha}$) denote the class of all α-excessive (resp. α-supermedian) functions (for Y) and as usual we shall drop α from our notation when $\alpha = 0$. Thus, for example, $\mathcal{E}$ is the class of all excessive functions. Note that $\underset{\sim}{S}^{\alpha} = \mathcal{S}^{\alpha} \cap \underset{\sim}{C}$ is unchanged by this. We shall need the following characterization of α-excessive functions by means of (P_t) in the following sections.

(5.6) PROPOSITION. *A function $f \in \underline{\underline{E}}^*_+$ is in $\mathcal{E}^{\alpha}$ if and only if $e^{-\alpha t} P_t f \uparrow f$ as $t \downarrow 0$.*

PROOF. Suppose first that $\alpha > 0$. If $g \in b\underline{\underline{E}}^*_+$, then using (3.19)

$$e^{-\alpha t} P_t U^{\alpha} g(x) = \int_0^{\infty} e^{-\alpha(t+s)} P_{t+s} g(x)\, ds = \int_t^{\infty} e^{-\alpha s} P_s g(x)\, ds \uparrow U^{\alpha} g(x)$$

as $t \downarrow 0$. If $f \in \mathcal{E}^\alpha$, by (2.9), there exists $(g_n) \subset b\underline{\underline{E}}^*_+$, with $U^\alpha g_n \uparrow f$. Therefore

$$e^{-\alpha t} P_t f = \lim_n e^{-\alpha t} P_t U^\alpha g_n \le \lim_n U^\alpha g_n = f ,$$

and since $e^{-\alpha t} P_t U^\alpha g_n$ increases as n increases and as t decreases one can interchange limits to obtain $e^{-\alpha t} P_t f \uparrow f$. If $f \in \mathcal{E}$, then $f \in \mathcal{E}^\alpha$ for each $\alpha > 0$ and letting $\alpha \downarrow 0$ in $e^{-\alpha t} P_t f \le f$ yields $P_t f \le f$. Again one can interchange limits to obtain $P_t f \uparrow f$. The converse is clear and left to the reader.

(5.7) DEFINITION. _A numerical function f on E is nearly Borel provided f is universally measurable and for each μ there exist $g, h \in \underline{\underline{E}}$ such that $g \le f \le h$ and the processes $(g \circ Y_t)$ and $(h \circ Y_t)$ are P^μ indistinguishable._

It is easy to see that the class of all $A \in \underline{\underline{E}}^*$ with 1_A nearly Borel forms a σ-algebra, $\underline{\underline{E}}^n$, and that f is nearly Borel if and only if it is $\underline{\underline{E}}^n$ measurable. Of course, $\underline{\underline{E}} \subset \underline{\underline{E}}^n \subset \underline{\underline{E}}^*$. If $D = E$, that is there are no branch points, then the explicit assumption that $f \in \underline{\underline{E}}^*$ in (5.7) is superfluous. If $g \in \underline{\underline{E}}$, then the process $(g \circ Y_t)$ is P^μ well measurable for each μ. Thus from the point of view of the general theory of processes, the fact that $f \in \underline{\underline{E}}^n$ has the important consequence that for each μ the process $(f \circ Y_t)$ is P^μ indistinguishable from a P^μ well measurable process which may, however, depend on μ.

(5.8) THEOREM. (i) _For each μ, $(\underline{\underline{G}}^\mu_t)_{t \ge 0}$ is right continuous._

(ii) _For each μ, $(Y_t, \underline{\underline{G}}^\mu_t, P^\mu)$ is strong Markov; that is, for each $(\underline{\underline{G}}^\mu_t)$ stopping time T, $f \in b\underline{\underline{E}}^*$, and $t \ge 0$ one has_

$$E^\mu[f(Y_{t+T}) 1_{\{T<\infty\}} \mid \underline{\underline{G}}^\mu_T] = P_t f(Y_T) 1_{\{T<\infty\}} .$$

(iii) _If f is α-excessive, then f is nearly Borel and almost surely $t \to f \circ Y_t$ is right continuous on $\mathbb{R}^+$ and has left limits on $\mathbb{R}^{++}$._

PROOF. Since $U^\alpha : \underset{\sim}{C} \to \underset{\sim}{C}$, (i) and (ii) are proved exactly as in BG-I-(8.11) and BG-I-(8.12) to which we refer the reader. We shall only prove (iii) here. In view of (2.8) it suffices to prove (iii) when $\alpha > 0$. Fix $\alpha > 0$ and let $\mathcal{H}$ be the collection of all $g \in b\underline{\underline{E}}$ such that almost surely $t \to U^\alpha g(Y_t)$ is right continuous and has left limits. Clearly $\mathcal{H}$ is a vector space and $\mathcal{H}$ contains $\underset{\sim}{C}$.

Suppose $(g_n) \subset \mathcal{K}^+$ and $0 \le g_n \uparrow g$ with g bounded. For each n and μ, $(e^{-\alpha t} U^\alpha g_n \circ Y_t)$ is a right continuous supermartingale with respect to P^μ and this family increases to $(e^{-\alpha t} U^\alpha g \circ Y_t)$. As a result by (4.1), $g \in \mathcal{K}$. Hence by the monotone class theorem $\mathcal{K} = b\underline{\underline{E}}$. Of course, $U^\alpha g \in b\underline{\underline{E}}$ if $g \in b\underline{\underline{E}}$. Next suppose $g \in b\underline{\underline{E}}^*_+$. Fix μ and let $\nu = \mu U^\alpha$, that is $\nu(B) = \int \mu(dx)\, U^\alpha(x, B)$ for all $B \in \underline{\underline{E}}$. Then there exist $g_1, g_2 \in b\underline{\underline{E}}_+$ with $g_1 \le g \le g_2$ and $\nu(g_2 - g_1) = 0$. Therefore

$$U^\alpha g_1 \circ Y_t \le U^\alpha g \circ Y_t \le U^\alpha g_2 \circ Y_t$$

for all t. But for each fixed t

$$E^\mu[U^\alpha g_2(Y_t) - U^\alpha g_1(Y_t)] = \int \mu(dx)\, P_t U^\alpha (g_2 - g_1)(x)$$

$$\le e^{\alpha t} \int \mu(dx)\, U^\alpha (g_2 - g_1)(x) = e^{\alpha t} \nu(g_2 - g_1) = 0 ,$$

and since $g_1, g_2 \in \mathcal{K}$ by what was proved above, it follows that $t \to U^\alpha g_1 \circ Y_t$ and $t \to U^\alpha g_2 \circ Y_t$ are P^μ indistinguishable. Since μ is arbitrary, this says that $U^\alpha g$ is nearly Borel and that almost surely $t \to U^\alpha g \circ Y_t$ is right continuous and has left limits. Finally if f is α-excessive, by (2.9) there exists $(g_n) \subset b\underline{\underline{E}}^*_+$ with $U^\alpha g_n \uparrow f$. Therefore f is nearly Borel and one more appeal to (4.1) completes the proof.

It follows from the strong Markov property (5.8-ii) that for $Z \in b\underline{\underline{G}}$ and T a $(\underline{\underline{G}}^\mu_t)$ stopping time one has

(5.9) $$E^\mu[Z \circ \theta_T\, 1_{\{T < \infty\}} | \underline{\underline{G}}^\mu_T] = E^{Y(T)}(Z)\, 1_{\{T < \infty\}} .$$

See, for example, BG-I-(8.6).

(5.10) REMARK. Suppose that f is a bounded α-excessive function. Then

$$e^{-\alpha t} P_t f(x) = e^{-\alpha t} E^x[f(Y_t)] ,$$

and letting $t \downarrow 0$ it follows from (5.6) and (5.8-iii) that $f(x) = E^x[f(Y_0)] = P_0 f(x)$. That is, $P_0 f = f$ for all bounded $f \in \mathcal{E}^\alpha$ and hence for all $f \in \mathcal{E}^\alpha$.

Because of the possible presence of branch points the process Y is not, in general, quasi-left-continuous. However the following result is a substitute for quasi-left-continuity under the present hypotheses, and it turns out to be quite adequate, as we shall see.

(5.11) THEOREM. Fix μ and let (T_n) be an increasing sequence of $(\underline{\underline{G}}_t^\mu)$ stopping times. Let $T = \sup T_n$ and $\Lambda = \{T < \infty;\ T_n < T$ for all $n\}$. If $f \in b\underline{\underline{E}}^*$, then

$$E^\mu\left[f \circ Y_T 1_{\{T<\infty\}} \mid \vee \underline{\underline{G}}^\mu_{T_n}\right] = f \circ Y_T 1_{\{T<\infty\}} 1_{\Lambda^c} + P_0 f(Y_{T-}) 1_\Lambda ,$$

where $\vee \underline{\underline{G}}^\mu_{T_n} = \sigma\left(\bigcup_{n \geq 1} \underline{\underline{G}}^\mu_{T_n}\right)$.

PROOF. First observe that $\Lambda \in \vee \underline{\underline{G}}^\mu_{T_n}$. Suppose that $\alpha > 0$ and $f \in \underset{\sim}{C}$, then using (5.5) we have

(5.12) $$E^\mu\left[\int_0^\infty e^{-\alpha s} f(Y_s)\,ds \mid \underline{\underline{G}}^\mu_t\right]$$

$$= \int_0^t e^{-\alpha s} f(Y_s)\,ds + E^\mu\left[\int_0^\infty e^{-\alpha(s+t)} f(Y_{s+t})\,ds \mid \underline{\underline{G}}^\mu_t\right]$$

$$= \int_0^t e^{-\alpha s} f(Y_s)\,ds + e^{-\alpha t} U^\alpha f(Y_t) .$$

Therefore $M_t = e^{-\alpha t} U^\alpha f(Y_t) + \int_0^t e^{-\alpha s} f(Y_s)\,ds$ is a right continuous version of the martingale on the left side of (5.12). Since M is bounded and $T_n \leq T$ we have (see D-V-T7, for example) $M_{T_n} = E^\mu[M_T \mid \underline{\underline{G}}^\mu_{T_n}]$. Of course, equations such as this and (5.12) hold almost surely P^μ. Letting $n \to \infty$ in this last equality results in

(5.13) $$\lim_n M_{T_n} = E^\mu\left[M_T \mid \vee \underline{\underline{G}}^\mu_{T_n}\right].$$

But

$$M_T = \int_0^T e^{-\alpha s} f(Y_s)\,ds + e^{-\alpha T} U^\alpha f(Y_T)$$

and the integral is $\vee \underset{=}{G}^{\mu}_{T_n}$ measurable, as is $e^{-\alpha T}$. Combining this with (5.13) gives

$$(5.14) \qquad E^{\mu}\left[U^{\alpha}f(Y_T)\, 1_{\{T<\infty\}} \mid \vee \underset{=}{G}^{\mu}_{T_n}\right] = \lim_n U^{\alpha}f(Y_{T_n})\, 1_{\{T<\infty\}}$$

$$= U^{\alpha}f(Y_T)\, 1_{\{T<\infty\}}\, 1_{\Lambda^c} + U^{\alpha}f(Y_{T-})\, 1_{\Lambda}$$

where we have used the continuity of $U^{\alpha}f$. Clearly $\alpha U^{\alpha}f \to P_0 f$ as $\alpha \to \infty$. Moreover $P_0 f(Y_T) = f(Y_T)$ if $T < \infty$ because $Y_T \in D$ if $T < \infty$ and $P_0 f(x) = f(x)$ if $x \in D$. Therefore if we multiply (5.14) by α and let $\alpha \to \infty$ we obtain

$$E^{\mu}\left[f(Y_T)\, 1_{\{T<\infty\}} \mid \vee \underset{=}{G}^{\mu}_{T_n}\right] = f(Y_T)\, 1_{\{T<\infty\}}\, 1_{\Lambda^c} + P_0 f(Y_{T-})\, 1_{\Lambda}$$

for all $f \in \underset{\sim}{C}$. A monotone class argument followed by a completion argument finishes the proof of Theorem 5.11.

(5.15) COROLLARY. With the same notation as in the statement of Theorem 5.11 the sets $\{Y_T = Y_{T-}\} \cap \Lambda$ and $\{Y_{T-} \in D\} \cap \Lambda$ are P^{μ} almost surely equal.

PROOF. Let $H = \{Y_{T-} \in D\} \cap \Lambda$. Since $Y_T \in D$ if $T < \infty$ one has $\{Y_T = Y_{T-}\} \cap \Lambda \subset H$. Moreover, $H \in \vee \underset{=}{G}^{\mu}_{T_n}$. Let $f, g \in \underset{\sim}{C}$. Then $g(Y_{T-})\, 1_H$ is $\vee \underset{=}{G}^{\mu}_{T_n}$ measurable and so by (5.11)

$$E^{\mu}[g(Y_{T-}) f(Y_T); H] = E^{\mu}[g(Y_{T-}) P_0 f(Y_{T-}); H] = E^{\mu}[g(Y_{T-})\, f(Y_{T-}); H] ,$$

where the last equality comes from the fact that $Y_{T-} \in D$ on H and $f = P_0 f$ on D. A monotone class argument now shows that for any $u \in b(\underset{=}{E} \otimes \underset{=}{E})$ one has

$$E^{\mu}[u(Y_{T-}, Y_T); H] = E^{\mu}[u(Y_{T-}, Y_{T-}); H] .$$

Letting u be the indicator of the diagonal in $E \times E$ we find that $Y_T = Y_{T-}$ almost surely P^{μ} on H. This establishes (5.15).

(5.16) REMARK. Since $Y_{T_n} \to Y_T$ on $\Lambda^c \cap \{T < \infty\}$, (5.15) implies that almost surely P^{μ}, $Y_{T_n} \to Y_T$ on $\{Y_{T-} \in D;\, T < \infty\}$. If there are no branch points, that is, $D = E$, then this is precisely the statement that the process Y is quasi-left-continuous.

(5.17) REMARK. If (T_n) is an increasing sequence of $(\underline{\underline{G}}_t)$ stopping times with limit T, then the conclusions of (5.11) and (5.15) hold for each initial measure μ.

(5.18) REMARK. If there are no branch points it is an easy consequence of (2.4), (2.7), and (5.8-iii) that if $f, g \in \mathcal{E}^\alpha$ then so is $f \wedge g$. See, for example, BG-II-(2.14). However, if B is not empty, this is false. In fact $f \in \mathcal{E}^\alpha$ does not even imply $f \wedge 1 \in \mathcal{E}^\alpha$. The reader is invited to furnish examples.

6. PROCESSES CONTINUED

The notation and assumptions are the same as in Section 5. The process Y constructed in (5.1) is called a Ray process, or perhaps better, the Ray process associated with the given Ray resolvent $(U^{\alpha})_{\alpha>0}$. In this section we shall develop some additional properties of such processes, and shall devote special attention to the properties of the σ-algebras $\{\underline{\underline{G}}_t^{\mu}\}$. However, in this section we shall make no explicit use of the fact that U^{α} sends $\underset{\sim}{C}$ into $\underset{\sim}{C}$; we shall use only the properties given in (5.1), (5.8), and (5.11).

(6.1) DEFINITION. *A point $x \in B$ is called a degenerate branch point if there exists $y \in E$ with $P_0(x, \cdot) = \epsilon_y$. The set of degenerate branch points is denoted by B_d.*

Since $P_0(x, \cdot)$ is carried by D for all x, $P_0(x, \cdot) = \epsilon_y$ with $y \in D$ if $x \in B_d$. If $x \in D$, then, of course, $P_0(x, \cdot) = \epsilon_x$. Let $\Phi\colon D \cup B_d \to D$ be the map such that $P_0(x, \cdot) = \epsilon_{\Phi(x)}$.

(6.2) LEMMA. *The set B_d is Borel and the map Φ is Borel measurable.*

PROOF. Since D is Borel and disjoint from B_d it suffices to show that $U = D \cup B_d$ is Borel and that Φ is Borel. One easily checks that a probability μ on $(E, \underline{\underline{E}})$ is unit mass at some point if and only if for each $n \geq 1$ there exists an open set G_n with $d(\overline{G}_n) \leq 1/n$ and $\mu(G_n) = 1$ for all n. Here $\overline{G}_n$ is the closure of G_n and $d(\overline{G}_n)$ is the diameter of $\overline{G}_n$. For each $n \geq 1$ let $\{G_k^n\}_{k\geq 1}$ be a countable base for the topology of E with $d(\overline{G}_k^n) \leq 1/n$ for all k. Then one easily verifies that

$$U = \bigcap_n \bigcup_k \{x\colon P_0(x, G_k^n) = 1\} \in \underline{\underline{E}},$$

and that if G is open $\Phi^{-1}(G) = \{x\colon P_0(x, G) = 1\} \cap U$. This establishes (6.2).

In general we shall use the terminology and notation in Dellacherie without special mention. It is convenient to introduce a point $\Delta \notin E$ and let $E_\Delta = E \cup \{\Delta\}$. We topologize E_Δ so that E as a subspace of E_Δ has its original topology and Δ is an isolated point of E_Δ. Then E_Δ is also a compact metric space. Finally we define $Y_\infty(w) = \Delta$ for all $w \in W$. The reason for all this is that we may write Y_T whenever T is a $(\underline{\underline{G}}_t^\mu)$ stopping time without always restricting it to $\{T < \infty\}$. Since $\{Y_T = \Delta\} = \{T = \infty\} \in \underline{\underline{G}}_{T-}^\mu$, the map $Y_T\colon W \to E_\Delta$ is $\underline{\underline{G}}_T^\mu$ measurable. We agree to extend any numerical function f on E to E_Δ by setting $f(\Delta) = 0$.

(6.3) LEMMA. Fix μ and let T be a $(\underline{\underline{G}}_t^\mu)$ stopping time. Let $\underline{\underline{H}} = \sigma(Y_{T+t};\ t \geq 0)$. Then $\underline{\underline{G}}^\mu = \sigma(\underline{\underline{H}} \cup \underline{\underline{G}}_{T-}^\mu \cup \underline{\underline{N}}^\mu)$ where, as in Section 5, $\underline{\underline{N}}^\mu = \{\Lambda \in \underline{\underline{G}}^\mu : P^\mu(\Lambda) = 0\}$.

PROOF. Since $\underline{\underline{S}} = \sigma(\underline{\underline{H}} \cup \underline{\underline{G}}_{T-}^\mu \cup \underline{\underline{N}}^\mu) \subset \underline{\underline{G}}^\mu$, it suffices to show that $f(Y_\tau)$ is $\underline{\underline{S}}$ measurable for each fixed $\tau \in \mathbb{R}^+$ and $f \in \underset{\sim}{C}$. If $\alpha > 0$ and $f \in \underset{\sim}{C}$, then

$$\int_0^\infty e^{-\alpha t} f(Y_t)dt = \int_0^T e^{-\alpha t} f(Y_t)dt + \int_T^\infty e^{-\alpha t} f(Y_t)dt$$

$$= \int_0^\infty e^{-\alpha t} f(Y_t)\, 1_{[0,T)}(t)dt + e^{-\alpha T} \int_0^\infty e^{-\alpha t} f(Y_{t+T})dt .$$

Since $f(Y_t)\, 1_{\{t<T\}}$ is $\underline{\underline{G}}_{T-}^\mu$ measurable, the first integral in the last displayed line is $\underline{\underline{G}}_{T-}^\mu$ measurable, while the second term is $\sigma(\underline{\underline{G}}_{T-}^\mu \cup \underline{\underline{H}})$ measurable because T is $\underline{\underline{G}}_{T-}^\mu$ measurable. Thus $\int_0^\infty e^{-\alpha t} f(Y_t)dt$ is $\underline{\underline{S}}$ measable for $\alpha > 0$ and $f \in \underset{\sim}{C}$. Hence by (3.12), $f(Y_\tau)$ is $\underline{\underline{S}}$ measurable for each $\tau \in \mathbb{R}^+$ establishing (6.3).

(6.4) PROPOSITION. Fix μ and let T be a previsible $(\underline{\underline{G}}_t^\mu)$ stopping time. Then $\underline{\underline{G}}_T^\mu = \underline{\underline{G}}_{T-}^\mu$ if and only if $P^\mu[Y_{T-} \in B - B_d;\ 0 < T < \infty] = 0$.

PROOF. Let (T_n) announce T. Then $\underline{\underline{G}}_{T-}^\mu = \vee\, \underline{\underline{G}}_{T_n}^\mu$ by D-III-T35. Therefore from (5.11) and the fact that $\Lambda = \{0 < T < \infty\}$ in the present situation, we obtain

(6.5) $E^{\mu}[f(Y_T)\,1_{\{T<\infty\}} \mid \underset{=}{G}^{\mu}_{T-}] = f(Y_0)\,1_{\{T=0\}} + P_0 f(Y_{T-})\,1_{\{0<T<\infty\}}$

for each $f \in b\underset{=}{E}$. Suppose first that $\underset{=}{G}^{\mu}_{T-} = \underset{=}{G}^{\mu}_{T}$. Then from (6.5), $f(Y_T) = P_0 f(Y_{T-})$ almost surely P^{μ} on $\{0<T<\infty\}$ for each $f \in b\underset{=}{E}$. Thus if $f, g \in b\underset{=}{E}$ we have P^{μ} almost surely on $\{0<T<\infty\}$,

(6.6) $$P_0(fg)(Y_{T-}) = f(Y_T)\,g(Y_T) = P_0 f(Y_{T-})\,P_0 g(Y_{T-}).$$

If $\lambda(dx) = P^{\mu}[Y_{T-} \in dx;\ 0<T<\infty]$ is the distribution of Y_{T-} on $\{0<T<\infty\}$ under P^{μ}, then (6.6) is equivalent to $P_0(fg) = (P_0 f)(P_0 g)$ almost everywhere λ. Using the separability of $\underset{\sim}{C}$ we see that there exists a subset E_0 of E with $\lambda(E_0) = 0$ such that if $x \notin E_0$, $P_0(fg)(x) = P_0 f(x)\,P_0 g(x)$ for all $f, g \in \underset{\sim}{C}$. But this implies that for each $x \notin E_0$ the measure $P_0(x, \cdot\,)$ must be unit mass at some point, that is, $x \in D \cup B_d$. As a result $Y_{T-} \in D \cup B_d$ almost surely P^{μ} on $\{0<T<\infty\}$, proving one of the assertions in (6.4).

For the converse suppose that $Y_{T-} \in D \cup B_d$ almost surely P^{μ} on $\{0<T<\infty\}$. Let $\Phi\colon D \cup B_d \to D$ be the Borel map defined above (6.2), that is, $P_0(x, \cdot\,) = \epsilon_{\Phi(x)}$. Now Y_{T-} is $\underset{=}{G}^{\mu}_{T-}$ measurable on $\{0<T<\infty\}$ because $Y_{T-} = \lim Y_{T_n}$ on $\{0<T<\infty\} \in \underset{=}{G}^{\mu}_{T-}$. Therefore, by (5.11), for $f, g \in b\underset{=}{E}$ we have

$$E^{\mu}[f(Y_T)\, g\circ\Phi(Y_{T-});\ 0<T<\infty] = E^{\mu}[P_0 f(Y_{T-})\, g\circ\Phi(Y_{T-});\ 0<T<\infty]$$

$$= E^{\mu}[f\circ\Phi(Y_{T-})\, g\circ\Phi(Y_{T-});\ 0<T<\infty],$$

where the last equality obtains because $Y_{T-} \in D \cup B_d$ almost surely and the definition of Φ. Therefore $E^{\mu}[F(Y_T, \Phi(Y_{T-})):\ 0<T<\infty] = E^{\mu}[F(\Phi(Y_{T-}), \Phi(Y_{T-})):\ 0<T<\infty]$ for all $F \in b(\underset{=}{E} \otimes \underset{=}{E})$. Hence $Y_T = \Phi(Y_{T-})$ almost surely P^{μ} on $\{0<T<\infty\}$ and so Y_T is $\underset{=}{G}^{\mu}_{T-}$ measurable on $\{0<T<\infty\}$. But $f(Y_0)\,1_{\{T=0\}} \in \underset{=}{G}^{\mu}_{T-}$ and $\{T=\infty\} \in \underset{=}{G}^{\mu}_{T-}$, and so Y_T is $\underset{=}{G}^{\mu}_{T-}$ measurable as a map from W to $(E_\Delta, \underset{=}{E}_\Delta)$.

Next suppose $Z = Z^1 \circ \theta_T\, Z^2$ where $Z^2 \in b\underset{=}{G}^{\mu}_{T-}$ and $Z^1 = \prod_{j=1}^{n} f_j(Y_{t_j})$ with $f_j \in b\underset{=}{E}$ for each j. Then $Z^1 \circ \theta_T \in b\underset{=}{H}$ where $\underset{=}{H}$ is the σ-algebra defined in the statement of Lemma 6.3. Now $E^{\mu}(Z \mid \underset{=}{G}^{\mu}_{T}) = Z^2 E^{Y(T)}(Z^1)$ which is $\underset{=}{G}^{\mu}_{T-}$ measurable because Y_T is measurable relative to $\underset{=}{G}^{\mu}_{T-}$ and $\underset{=}{E}_\Delta$.

Therefore

$$E^{\mu}[Z \mid \underline{\underline{G}}^{\mu}_{T-}] = E^{\mu}[E^{\mu}(Z \mid \underline{\underline{G}}^{\mu}_{T}) \mid \underline{\underline{G}}^{\mu}_{T-}] = Z^2 E^{Y(T)}(Z^1) = E^{\mu}[Z \mid \underline{\underline{G}}^{\mu}_{T}]$$

for all Z of the above form. Using (6.3) and the monotone class theorem this yields $E^{\mu}(Z \mid \underline{\underline{G}}^{\mu}_{T-}) = E^{\mu}(Z \mid \underline{\underline{G}}^{\mu}_{T})$ for all $Z \in b\underline{\underline{G}}^{\mu}$, and since $\underline{\underline{G}}^{\mu}_{T-} \subset \underline{\underline{G}}^{\mu}_{T} \subset \underline{\underline{G}}^{\mu}$ we see that $\underline{\underline{G}}^{\mu}_{T-} = \underline{\underline{G}}^{\mu}_{T}$, completing the proof of (6.4).

The following corollary is an immediate consequence of (6.4) and D-III-T51.

(6.7) COROLLARY. *If $B = B_d$, in particular, if B is empty, the family $(\underline{\underline{G}}^{\mu}_{t})$ is quasi-left-continuous for each μ.*

(6.8) PROPOSITION. *For each μ the set $H = \{(t, w): t > 0,\ Y_{t-}(w) \in B\}$ is equal to a countable union of disjoint graphs of $(\underline{\underline{G}}^{\mu}_{t})$ previsible stopping times.*

PROOF. Fix μ. Then the set H is previsible since Y_{t-} is left continuous. Because $Y_t \in D$

$$H \subset \{(t, w): Y_t(w) \neq Y_{t-}(w)\},$$

and this last set is a countable union of graphs of stopping times. See D-IV-T30. The proof of (6.8) is now completed by an appeal to D-IV-T17.

The next result sharpens the defining property of a nearly Borel function.

(6.9) PROPOSITION. *Let f be nearly Borel. Then for each μ there exist $h, g \in \underline{\underline{E}}$ with $g \leq f \leq h$ such that the processes $(h \circ Y_t)$ and $(g \circ Y_t)$, and also the processes $(h \circ Y_{t-})$ and $(g \circ Y_{t-})$ are P^{μ} indistinguishable.*

PROOF. Let μ be fixed and all statements refer to the system $(W, \underline{\underline{G}}^{\mu}, \underline{\underline{G}}^{\mu}_{t}, P^{\mu})$. There exist $g_1, h_1 \in \underline{\underline{E}}$ with $g_1 \leq f \leq h_1$ such that $(g_1 \circ Y_t)$ and $(h_1 \circ Y_t)$ are P^{μ} indistinguishable. Let $J = \{(t, w): 0 < t < \infty,\ Y_t(w) \neq Y_{t-}(w)\}$. Then $J = \bigcup_n [[T_n]]$ where each T_n is a stopping time. See D-IV-T30. Let $\nu_n(dx) = P^{\mu}[Y_{T_n-} \in dx;\ T_n < \infty]$ and $\nu = \sum 2^{-n} \nu_n$. Since f is universally measurable there exist $g_2, h_2 \in \underline{\underline{E}}$ with $g_2 \leq f \leq h_2$ and $\nu(\{g_2 < h_2\}) = 0$.

Finally define $g = g_1 \vee g_2$ and $h = h_1 \wedge h_2$. Clearly $(g \circ Y_t)$ and $(h \circ Y_t)$ are P^μ indistinguishable. But

$$\{(t,w):\ g \circ Y_{t-}(w) \neq h \circ Y_{t-}(w);\ 0 < t < \infty\}$$

$$\subset \bigcup_n \{(t,w):\ g \circ Y_{T_n-}(w) \neq h \circ Y_{T_n-}(w);\ T_n(w) < \infty\}$$

$$\cup \{(t,w):\ g \circ Y_t(w) \neq h \circ Y_t(w);\ 0 < t < \infty\},$$

and each of these sets is P^μ evanescent.

7. CHARACTERIZATION OF PREVISIBLE STOPPING TIMES

In this section we shall characterize previsible $(\underline{\underline{G}}_t^\mu)$ stopping times, and also the totally inaccessible part of a $(\underline{\underline{G}}_t^\mu)$ stopping time. The assumptions are the same as in Section 6, and again, in this section, we shall make no explicit use of the fact that U^α sends $\underset{\sim}{C}$ into $\underset{\sim}{C}$.

Recall from D-V-D11 that a positive uniformly integrable right continuous supermartingale $X = (X_t)$ over a system $(\Omega, \underline{\underline{F}}, \underline{\underline{F}}_t, P)$ is called a potential if $\lim_{t \to \infty} X_t = 0$, or equivalently, $\lim_{t \to \infty} E(X_t) = 0$. Such a supermartingale, X, is regular if $E(X_T) = E(X_{T-})$ for all previsible stopping times T. See D-V-D51. It is of class (D) if the family of random variables $\{X_T: \text{ T a stopping time}\}$ is uniformly integrable (D-V-D13). It is immediate that if X is of class (D), then X is regular if and only if $\lim_n E(X_{T_n}) = E(X_T)$ for all previsible T and any sequence (T_n) announcing T. If X is a regular potential of class (D), then there exists a continuous, integrable, increasing, adapted process A such that $X_T = E(A_\infty | \underline{\underline{F}}_T) - A_T$ for all stopping times T. See D-V-T52 and D-V-T47. It follows at once from this representation that if X is a regular potential of class (D), then $\lim_n E(X_{T_n}) = E(X_T)$ whenever (T_n) is an increasing sequence of stopping times with limit T.

Having recalled these facts from the general theory of processes, let us return to the Ray process Y. If g is a bounded α-excessive function and $\alpha > 0$, then it is immediate that $(e^{-\alpha t} g(Y_t), \underline{\underline{G}}^\mu, \underline{\underline{G}}_t^\mu, P^\mu)$ is a bounded (and hence class (D)) potential for each μ.

(7.1) DEFINITION. If $g \in b\mathcal{E}^\alpha$ with $\alpha > 0$, then g is μ regular if the potential $(e^{-\alpha t} g(Y_t))$ is P^μ regular. We say g is regular if it is μ regular for each μ.

In view of the above remarks it is clear that a $g \in b\mathcal{E}^{\alpha}$ with $\alpha > 0$ is μ regular if and only if $E^{\mu}\left(e^{-\alpha T_n} g \circ Y_{T_n}\right) \to E^{\mu}(e^{-\alpha T} g \circ Y_T)$ whenever (T_n) is an increasing sequence of $(\underline{\underline{G}}_t^{\mu})$ stopping times with limit T.

(7.2) PROPOSITION. If $f \in b\underline{\underline{E}}_+^*$ and $\alpha > 0$, then $g = U^{\alpha} f$ is regular.

PROOF. Fix μ and let (T_n) be an increasing sequence of $(\underline{\underline{G}}_t^{\mu})$ stopping times with limit T. Then

$$E^{\mu}\left(e^{-\alpha T_n} U^{\alpha} f \circ Y_{T_n}\right) = E^{\mu} \int_{T_n}^{\infty} e^{-\alpha t} f(Y_t)\, dt$$

$$\to E^{\mu} \int_{T}^{\infty} e^{-\alpha t} f(Y_t) dt = E^{\mu}(e^{-\alpha T} U^{\alpha} f \circ Y_T) ,$$

proving (7.2).

In the next result we are going to compare $g(Y_{t-}) = g(\lim_{s \uparrow t} Y_s)$ and $g(Y_t)_- = \lim_{s \uparrow t} g(Y_s)$. This last expression exists almost surely by (5.8).

(7.3) THEOREM. Let g be a bounded α-excessive function with $\alpha > 0$ and let μ be fixed.

(i) P^{μ} almost surely $g(Y_t)_- \geq g(Y_{t-})$ for all $t > 0$.

(ii) g is μ regular if and only if P^{μ} almost surely $g(Y_t)_- = g(Y_{t-})$ for all $t > 0$.

REMARKS. (i) We emphasize that (7.3-i) states that for each w outside of a set of P^{μ} measure zero one has $g(Y_t(w))_- \geq g(Y_{t-}(w))$ for each $t > 0$. We shall show in (7.18) that there actually is equality if $Y_{t-}(w) \in D$ and $Y_{t-}(w) \neq Y_t(w)$.

(ii) It follows from (7.3-ii) that the concept of regularity is independent of α.

(iii) (7.3-i) holds for arbitrary α-excessive functions, $\alpha \geq 0$. To see this suppose first that $g \in \mathcal{E}^{\alpha}$ with $\alpha > 0$. Then by (2.7) there exists a sequence (g_n) of bounded α-excessive functions increasing to g. Thus, by (7.3-i), $g(Y_t)_- \geq g_n(Y_t)_- \geq g_n(Y_{t-})$, and letting $n \to \infty$ we obtain $g(Y_t)_- \geq g(Y_{t-})$. The result then holds for excessive functions since any such function is α-excessive for all $\alpha > 0$.

PROOF. Let

$$U = \{(t,w):\ t > 0,\ g(Y_t(w))_- < g(Y_{t-}(w))\}\ .$$

Then (7.3-i) is equivalent to the statement that U is P^μ evanescent. Since g is nearly Borel, the processes $(g(Y_{t-}))$ and $(g(Y_t)_-)$ are previsible (actually only P^μ indistinguishable from previsible processes), and hence U is previsible. If U is not P^μ evanescent, then using the section theorem (D-IV-T10) one may find a previsible stopping time T with $[[T]] \subset U$ and $P^\mu(T<\infty) > 0$. Let (T_n) announce T. By the supermartingale inequality

$$e^{-\alpha T_n} g(Y_{T_n}) \geq E^\mu\left[e^{-\alpha T} g(Y_T) \mid \underline{\underline{G}}^\mu_{T_n}\right],$$

and letting $n \to \infty$ and using (5.11) one finds $(T > 0$ since $[[T]] \subset U)$

$$e^{-\alpha T} g(Y_T)_- \geq E^\mu\left[e^{-\alpha T} g(Y_T) \mid \vee\, \underline{\underline{G}}^\mu_{T_n}\right] = e^{-\alpha T} P_0 g(Y_{T-})\ .$$

But g is α-excessive and so $g = P_0 g$ according to (5.10). Hence $g(Y_T)_- \geq g(Y_{T-})$ almost surely P^μ on $\{T < \infty\}$. But this contradicts the fact that $[[T]] \subset U$, establishing (7.3-i).

Coming to (7.3-ii) let

$$V = \{(t,w):\ t > 0,\ g(Y_t(w))_- > g(Y_{t-}(w))\}\ .$$

In light of (7.3-i) in order to establish (7.3-ii) we must show that V is P^μ evanescent if and only if g is μ regular. Suppose first that g is μ regular. As before V is previsible and so if it is not evanescent we can find a previsible stopping time T with $[[T]] \subset V$ and $P^\mu(T < \infty) > 0$. By the definition of regularity

$$E^\mu[e^{-\alpha t} g(Y_T)_-] = E^\mu[e^{-\alpha T} g(Y_T)]\ ,$$

and by (7.3-i), $g(Y_T)_- \geq g(Y_{T-})$ almost surely P^μ on $\{T < \infty\}$. But from (5.11) and $g = P_0 g$,

(7.5) $$E^\mu[e^{-\alpha T} g(Y_T)] = E^\mu[e^{-\alpha T} P_0 g(Y_{T-})] = E^\mu[e^{-\alpha T} g(Y_{T-})]\ ,$$

and combining these facts one finds that $g(Y_T)_- = g(Y_{T-})$ almost surely P^μ on $\{T < \infty\}$. This contradicts $[[T]] \subset V$ and V is P^μ evanescent.

Conversely suppose that g is a bounded α-excessive function and that V is P^μ evanescent. Let T be a previsible stopping time. Then because both U and V are P^μ evanescent

$$E^\mu[e^{-\alpha T} g(Y_T)_-] = E^\mu[e^{-\alpha T} g(Y_{T-})] ,$$

and combining this with (7.5) we see that g is μ regular completing the proof of (7.3).

We come now to the main result of this section. Recall that if T is a $(\underline{\underline{G}}^\mu_t)$ stopping time and $A \in \underline{\underline{G}}^\mu_T$, then $T_A = T$ on A, $T_A = \infty$ on A^c is again a $(\underline{\underline{G}}^\mu_t)$ stopping time.

(7.6) THEOREM. *Fix μ and let T be a $(\underline{\underline{G}}^\mu_t)$ stopping time.*

(i) If $Y_T = Y_{T-}$ almost surely P^μ on $\{0 < T < \infty\}$, then T is previsible and $\underline{\underline{G}}^\mu_T = \underline{\underline{G}}^\mu_{T-}$.

(ii) The totally inaccessible part of T is T_A where $A = \{0 < T < \infty, Y_{T-} \in D, Y_T \neq Y_{T-}\}$.

PROOF. To establish (7.6-i) it suffices to show that T is previsible because the hypothesis implies $P^\mu[Y_{T-} \in B, 0 < T < \infty] = 0$, and so $\underline{\underline{G}}^\mu_T = \underline{\underline{G}}^\mu_{T-}$ follows from (6.4) once we know that T is previsible. According to D-V-T43, in order to show that T is previsible it suffices to show that $E^\mu(M_{T-}) = E^\mu(M_T)$ for all bounded right continuous martingales $M = (M_t)$ over the system $(W, \underline{\underline{G}}^\mu, \underline{\underline{G}}^\mu_t, P^\mu)$. If M is such a martingale, then $M_\infty = \lim_{t\to\infty} M_t$ exists almost surely and $M_t = E^\mu(M_\infty | \underline{\underline{G}}^\mu_t)$. I claim that it certainly suffices to show $M_T = M_{T-}$ almost surely P^μ for all right continuous martingales $M_t = E^\mu(M_\infty | \underline{\underline{G}}^\mu_t)$ as M_∞ ranges over a dense subset of $L^2(P^\mu) = L^2(W, \underline{\underline{G}}^\mu, P^\mu)$. To see this suppose $M^n_\infty \to M_\infty$ in $L^2(P^\mu)$ and let (M^n_t) and (M_t) be right continuous versions of the martingales $E^\mu(M^n_\infty | \underline{\underline{G}}^\mu_t)$ and $E^\mu(M_\infty | \underline{\underline{G}}^\mu_t)$ respectively. By passing to a subsequence we may assume that $\sum_n E^\mu(|M^n_\infty - M_\infty|^2) < \infty$. But then by Doob's inequality (see V-T22 of [8])

$$E^\mu\left(\sup_{0 \le t \le \infty} |M^n_t - M_t|^2\right) \le 4\, E^\mu(|M^n_\infty - M_\infty|^2) ,$$

and by Tchebychev's inequality for each $\varepsilon > 0$

$$P^{\mu}\left(\sup_{0 \le t \le \infty} |M_t^n - M_t| > \varepsilon \right) \le \varepsilon^{-2} E^{\mu}\left(\sup_{0 \le t \le \infty} |M_t^n - M_t|^2 \right).$$

Combining these estimates with the Borel-Cantelli lemma we see that almost surely P^{μ}, $M_t^n \to M_t$ uniformly on $[0, \infty]$. Consequently if each M^n is continuous at T so is M.

Thus it remains to choose a convenient family (M_{∞}) dense in $L^2(P^{\mu})$ and verify $M_T = M_{T-}$ for this family. Since $M_T = M_{T-}$ is preserved under the formation of linear combinations it suffices to find a family (M_{∞}) that is total in $L^2(P^{\mu})$, that is, whose linear span is dense. To this end we shall show that the set of all finite products of the form

$$(7.7) \qquad M_{\infty} = \prod_{j=1}^{n} \int_0^{\infty} e^{-\alpha_j t} f_j(Y_t)\, dt$$

where $\alpha_j > 0$, $f_j \in \underset{\sim}{C}^+$, $n \ge 1$ is total in $L^2(P^{\mu})$. Note first that each M_{∞} of the form (7.7) is in $b\underline{\underline{G}}^0$. Let $\mathcal{K}$ be the vector space generated by random variables of the form (7.7). Then $1 \in \mathcal{K}$ and $\mathcal{K} \subset b\underline{\underline{G}}^0 \subset L^2(P^{\mu})$. Let $\overline{\mathcal{K}}$ be the closure of $\mathcal{K}$ in $L^2(P^{\mu})$. Suppose that we can show that $\overline{\mathcal{K}}$ contains all random variables of the form

$$(7.8) \qquad \prod_{j=1}^{n} e^{-t_j} f_j(Y_{t_j})$$

with $f_j \in \underset{\sim}{C}$, $t_j \ge 0$, and $n \ge 1$. Then by the monotone class theorem, see BG-O-(3.6), $\overline{\mathcal{K}}$ contains $b\underline{\underline{G}}^0$ and hence $\overline{\mathcal{K}} = L^2(P^{\mu})$. Thus to show elements of the form (7.7) form a total set it suffices to show that $\overline{\mathcal{K}}$ contains all elements of the form (7.8). Starting from (7.7) it follows by elementary algebra that $\mathcal{K}$ contains all products of the form

$$(7.9) \qquad \prod_{j=1}^{n} \int_0^{\infty} e^{-t} \phi_j(t) f_j(Y_t)\, dt$$

where each $f_j \in \underset{\sim}{C}$ and each ϕ_j is a polynomial in e^{-t}. Since $\overline{\mathcal{K}}$ is closed under uniform limits, it follows from the Stone-Weierstrass theorem that $\overline{\mathcal{K}}$ contains all products of the form (7.9) where now the ϕ_j's are continuous and vanish at infinity on $\mathbb{R}^+$. By the argument used in the proof of (3.12) we can find a sequence of elements of the form (7.9) which converge boundedly, and hence in $L^2(P^\mu)$, to each element of the form (7.8). This establishes the fact that elements of the form (7.7) are total in $L^2(P^\mu)$.

To complete the proof of (7.6-i) we must show that if (M_t) is a right continuous version of the martingale $E^\mu(M_\infty | \underline{\underline{G}}^\mu_t)$ where M_∞ is of the form (7.7), then $M_T = M_{T-}$ almost surely P^μ. Since $M_\infty = \lim_{t\to\infty} M_t = M_{\infty-}$ almost surely P^μ, this amounts to showing that $M_T = M_{T-}$ almost surely P^μ on $\{T < \infty\}$.

In the course of the argument we shall need the following lemma.

(7.10) LEMMA. Let $g \in b\underline{\underline{E}}^*$ be such that for a fixed $\alpha > 0$, $t^{-1}(g - e^{-\alpha t} P_t g)$ is uniformly bounded for $t > 0$ and $x \in E$ and suppose that $t^{-1}(g - e^{-\alpha t} P_t g) \to f$ as $t \downarrow 0$. Then $U^\alpha f = g$.

PROOF. Clearly the hypotheses imply that $f \in b\underline{\underline{E}}^*$ and that $e^{-\alpha t} P_t g \to g$ as $t \downarrow 0$. Since $\alpha > 0$ we have by the bounded convergence theorem

$$
\begin{aligned}
U^\alpha f &= \lim_{t\downarrow 0} U^\alpha [t^{-1}(g - e^{-\alpha t} P_t g)] \\
&= \lim_{t\downarrow 0} t^{-1}\left[\int_0^\infty e^{-\alpha s} P_s g\, ds - e^{-\alpha t}\int_0^\infty e^{-\alpha s} P_{t+s} g\, ds\right] \\
&= \lim_{t\downarrow 0} t^{-1}\int_0^t e^{-\alpha s} P_s g\, ds = g ,
\end{aligned}
$$

proving (7.10).

We return now to the proof of (7.6-i). Suppose first $M_\infty = \int_0^\infty e^{-\alpha t} f(Y_t)dt$ with $\alpha > 0$ and $f \in \underset{\sim}{C}^+$. Then

$$
M_t = \int_0^t e^{-\alpha s} f(Y_s)ds + e^{-\alpha t} U^\alpha f(Y_t) \tag{7.11}
$$

is a right continuous version of $E^\mu(M_\infty | \underline{\underline{G}}^\mu_t)$. The integral in (7.11) is continuous in t while $U^\alpha f$ is regular by (7.2). Consequently by (7.3-ii), $U^\alpha f(Y_T)_- = U^\alpha f(Y_{T-})$ almost surely P^μ on $\{T < \infty\}$. But using the hypothesis of (7.6-i) for the first time we see that $M_T = M_{T-}$ almost surely P^μ on $\{T < \infty\}$. We now proceed to the general case, and for this we need some notation. If $\alpha_1, \ldots, \alpha_n > 0$ and $f_1, \ldots, f_n \in \underset{\sim}{C}^+$ define

$$I^n(s) = I^n(\alpha_1, f_1; \ldots; \alpha_n, f_n; s) = \prod_{j=1}^{n} \int_0^s e^{-\alpha_j t} f_j(Y_t)dt \; ,$$

(7.12)

$$J^n(x) = J^n(\alpha_1, f_1; \ldots; \alpha_n, f_n; x) = E^x \prod_{j=1}^{n} \int_0^\infty e^{-\alpha_j t} f_j(Y_t)\, dt \; .$$

Thus $I^n(s)$ is a continuous $(\underset{=}{G}{}_t^0)$ adapted process and $J^n(x)$ is a bounded Borel function on E. Note that $J^1(\alpha_1, f_1; x) = U^{\alpha_1} f_1(x)$.

We shall show next that each J^n is a potential $U^\beta g$ for appropriate $\beta > 0$ and $g \in b\underset{=}{E}^+$. Given $\alpha_1, \ldots, \alpha_n > 0$ and $f_1, \ldots, f_n \in \underset{\sim}{C}^+$, let $\beta = \alpha_1 + \cdots + \alpha_n$. Then

$$J^n(x) - e^{-\beta t} P_t J^n(x) = E^x\left(\prod_{j=1}^{n} \int_0^\infty e^{-\alpha_j s} f_j(Y_s)ds - \prod_{j=1}^{n} \int_t^\infty e^{-\alpha_j s} f_j(Y_s)\, ds \right) .$$

Using the identity

$$a_1 a_2 \cdots a_n - b_1 b_2 \cdots b_n = (a_1 - b_1) a_2 \cdots a_n$$
$$+ b_1(a_2 - b_2)a_3 \cdots a_n + \cdots + b_1 \cdots b_{n-1}(a_n - b_n) ,$$

one easily sees that $t^{-1}(J^n - e^{-\beta t} P_t J^n)$ is bounded in t and x and approaches $P_0 g$ as $t \to 0$ where

$$g = \sum_{j=1}^{n} f_j \, J^{n-1}(\alpha_1, f_1; \cdots; \hat{\alpha}_j, \hat{f}_j; \cdots; \alpha_n, f_n; \cdot \;)$$

in which the hat " ^ " indicates that these quantities have been omitted. Therefore by (7.10), $J^n = U^\beta P_0 g = U^\beta g$ because $U^\beta P_0 = U^\beta$. Since $g \in b\underset{=}{E}^+$, $J^n = U^\beta g$ is regular, and so as before $t \to J^n(Y_t)$ is right continuous and continuous at T. Finally if M_∞ is of the form (7.7), then writing each $\int_0^\infty$ as $\int_0^t + \int_t^\infty$ one easily checks using the notation of (7.12) that

(7.13)

$$E^{\mu}(M_\infty | \underline{\underline{G}}_t^{\mu}) = I^n(\alpha_1, f_1; \cdots ; \alpha_n, f_n; t)$$

$$+ \sum_{j=1}^{n} I^{n-1}(\alpha_1, f_1; \cdots ; \hat{\alpha}_j, \hat{f}_j; \cdots ; \alpha_n, f_n; t)\, J^1(\alpha_j, f_j; Y_t)\, e^{-\alpha_j t}$$

$$+ \sum_{i<j} I^{n-1}(\alpha_1, f_1; \cdots ; \hat{\alpha}_i, \hat{f}_i; \cdots ; \hat{\alpha}_j, \hat{f}_j; \cdots ; \alpha_n, f_n; t)$$

$$\times J^2(\alpha_i, f_i, \alpha_j, f_j; Y_t)\, e^{-(\alpha_i+\alpha_j)t} + \cdots + J^n(\alpha_1, f_1; \cdots ; \alpha_n, f_n; Y_t)$$

$$\times e^{-(\alpha_1 + \cdots + \alpha_n)t} ,$$

where again the hat " ^ " indicates quantities that have been omitted. Since each $I^k(t)$ is continuous and each $J^k = U^{\beta_k} g_k$ for appropriate $\beta_k > 0$ and $g_k \in b\underline{\underline{E}}^+$, it follows that the right side of (7.13) is a right continuous version of $E^{\mu}(M_\infty | \underline{\underline{G}}_t^{\mu})$ which is continuous at T almost surely P^{μ} on $\{T < \infty\}$. This completes the proof of (7.6-i).

We turn now to (7.6-ii). By considering T_A and T_{A^c} separately, the statement of (7.6-ii) is equivalent to the following two statements about a $(\underline{\underline{G}}_t^{\mu})$ stopping time T.

(7.14) *If* $T > 0$, $Y_{T-} \in D$, *and* $Y_{T-} \neq Y_T$ *almost surely* P^{μ} *on* $\{T < \infty\}$, *then* T *is totally inaccessible.*

(7.15) *If* P^{μ} *almost surely on* $\{T < \infty\}$ *one has either* $T = 0$ *or* $Y_{T-} \in B$ *or* $Y_T = Y_{T-}$, *then* T *is accessible.*

For the proof of (7.14) suppose that R is previsible and that $P^{\mu}[R = T < \infty] > 0$. Let (R_n) announce R. According to (5.15) one has $\{Y_R = Y_{R-};\, R < \infty\} = \{Y_{R-} \in D,\, R < \infty\}$ almost surely P^{μ}. Consequently

$$0 < P^{\mu}[R = T < \infty] = P^{\mu}[Y_{T-} \in D;\, R = T < \infty] \le P^{\mu}[Y_{T-} \in D;\, Y_T = Y_{T-};\, T < \infty] ,$$

contradicting the hypothesis on T. Therefore T is totally inaccessible.

Coming to (7.15), let $J = \{T = 0\}$, $K = \{0 < T < \infty ; Y_{T-} \in B\}$ and $L = \{0 < T < \infty; Y_T = Y_{T-}\}$. Since $T = T_J \wedge T_K \wedge T_L$ it suffices to show that each of T_J, T_K, and T_L is accessible. For T_J this is clear because 0 is previsible. By (7.6-i), T_L is even previsible. For T_K observe that

$$[[T_K]] \subset \{(t, w) ; t > 0, \ Y_{t-}(w) \in B\} ,$$

and that by (6.8) this last set is a countable union of graphs of previsible stopping times. Therefore T_K is accessible and (7.15) is established, and Theorem 7.6 finally is proved.

The techniques used in the proof of (7.6) are useful in many other situations. We illustrate this by giving a characterization of well measurable processes taken from [5]. Let $\underline{\underline{A}}$ be the σ-algebra on $\mathbb{R}^+ \times W$ that is generated by processes of the form $(t, w) \to X_t(w) f(Y_t(w))$ where $X = (X_t)$ is a continuous $(\underline{\underline{G}}_t^0)$ adapted process and $f \in \underset{\sim}{C}$. Of course, one obtains the same σ-algebra $\underline{\underline{A}}$ if one allows f to range over $b\underline{\underline{E}}$. If μ is fixed, then an $\underline{\underline{A}}$ measurable process is certainly well measurable over $(W, \underline{\underline{G}}^\mu, \underline{\underline{G}}_t^\mu, P^\mu)$.

(7.16) PROPOSITION. <u>If</u> $Z = (Z_t)$ <u>is well measurable over</u> $(W, \underline{\underline{G}}^\mu, \underline{\underline{G}}^\mu, P^\mu)$, <u>then</u> Z <u>is</u> P^μ <u>indistinguishable from an</u> $\underline{\underline{A}}$ <u>measurable process</u>.

PROOF. If Z is a bounded $\underline{\underline{R}}^+ \otimes \underline{\underline{G}}^\mu$ measurable process, let ${}^1Z$ denote the well measurable projection of Z. All statements are relative to $(W, \underline{\underline{G}}^\mu, \underline{\underline{G}}_t^\mu, P^\mu)$. Let $\mathcal{K}$ be the collection of all such Z for which ${}^1Z$ is indistinguishable from an $\underline{\underline{A}}$ measurable process. To prove (7.16) it suffices to show that $\mathcal{K}$ contains all bounded $\underline{\underline{R}}^+ \otimes \underline{\underline{G}}^\mu$ measurable processes. But $\mathcal{K}$ is a vector space containing 1 and closed under uniform limits and also bounded monotone increasing limits. See the proof of D-V-T14. Hence it suffices to show that $\mathcal{K}$ contains all processes of the form

$$Z_t(w) = 1_{[a,b]}(t)\, M(w) \tag{7.17}$$

where $M \in b\underline{\underline{G}}^\mu$ and $0 \le a < b < \infty$. In this case, ${}^1Z_t(w) = 1_{[a,b]}(t)\, M_t(w)$ where (M_t) is a right continuous version of the martingale $E^\mu(M \mid \underline{\underline{G}}_t^\mu)$. If M is of the form (7.7), then (7.12) and (7.13) show that ${}^1Z$ is indistinguishable from an $\underline{\underline{A}}$ measurable process. But using the fact that elements of the form (7.7) form a total set in $L^2(P^\mu)$ and arguing as in the first paragraph of the proof (7.6) we see that ${}^1Z$ is indistinguishable from an $\underline{\underline{A}}$ measurable process for all Z of the form (7.17). This establishes (7.16).

We close this section by sharpening the result given in (7.3-i).

(7.18) COROLLARY. Fix μ and let $g \in \mathcal{E}^\alpha$. Then P^μ almost surely $g(Y_t)_- = g(Y_{t-})$ for all $t > 0$ such that $Y_{t-} \in D$ and $Y_t \neq Y_{t-}$.

PROOF. Let $\Lambda = \{(t, w): t > 0, g(Y_t(w))_- \neq g(Y_{t-}(w))\}$. Then Λ is (P^μ indistinguishable from) a previsible set and is contained in

$$\{(t, w): g(Y_t(w))_- \neq g(Y_t(w))\} \cup \{(t, w): Y_{t-}(w) \neq Y_t(w)\}.$$

But each of these last sets is a countable union of graphs of stopping times and so by D-IV-T17, Λ is a countable union of graphs of previsible stopping times. Similarly (7.6-ii) implies that $\Gamma = \{(t, w): t > 0, Y_{t-}(w) \in D, Y_{t-}(w) \neq Y_t(w)\}$ is a countable union of graphs of totally inaccessible stopping times. Consequently $\Lambda \cap \Gamma$ is P^μ evanescent proving (7.18).

(7.19) REMARK. In the following discussion μ is fixed and all equalities hold P^μ almost surely and all statements are relative to $(W, \underline{\underline{G}}^\mu, \underline{\underline{G}}_t^\mu, P^\mu)$. Let g be an α-excessive function and consider the following statements:

(a) $g(Y_t)_- = g(Y_{t-})$ for all $t > 0$.

(b) $t \to g(Y_t)$ is continuous wherever $t \to Y_t$ is continuous.

If g is bounded and $\alpha > 0$ then (a) is equivalent to g being μ regular according to (7.3-ii). Observe that (a) implies (b) because if $Y_t = Y_{t-}$, then $g(Y_t) = g(Y_{t-}) = g(Y_t)_-$. Conversely if Y is quasi-left-continuous (b) implies (a). To see this note that $\Lambda = \{(t, w): t > 0, g(Y_{t-}(w)) \neq g(Y_t(w))_-\}$ is previsible. If T is a previsible stopping time with $[[T]] \subset \Lambda$, then the quasi-left-continuity of Y implies that $Y_T = Y_{T-}$ on $\{T < \infty\}$ and so if (b) holds $g(Y_T)_- = g(Y_T) = g(Y_{T-})$ on $\{T < \infty\}$. Hence $P^\mu(T < \infty) = 0$ and Λ is evanescent, that is, (a) holds. In BG condition (b) is used as the definition of regularity.

8. SOME TOPOLOGY AND MEASURE THEORY

In the first part of this section we introduce a class of topological spaces (U-spaces) that will be of importance in later sections. We then develop the properties of these spaces that will be of immediate use. Some deeper facts about them are presented in Section 14. In (8.7) we state the famous theorem of Lusin for ease of future reference. The remainder of the section contains some standard facts from measure theory that we shall need later. The reader may prefer to skip the second part of this section and go directly from (8.7) to Section 9, referring back to the latter part of this section only as needed.

(8.1) DEFINITION. *A topological space E is a U-space (resp. Lusinien) if it is homeomorphic to a universally measurable (resp. Borel) subspace of a compact metric space.*

Here a universally measurable (resp. Borel) subspace of a topological space means a universally measurable (resp. Borel) subset with the subspace topology. It is immediate that a Lusinien space is a U-space, and that a U-space is metrizable and second countable. Let E be a U-space and let $\hat{E}$ be a compact metric space containing a homeomorphic image of E. For ease of exposition we identify E with a universally measurable subspace of $\hat{E}$. Fix a metric d on $\hat{E}$ compatible with the topology of $\hat{E}$ so that the topology of the metric space (E, d) is precisely the topology of $\hat{E}$. Let $\mathbf{C}_b = \mathbf{C}_b(E)$ denote the bounded real valued continuous functions on E and let $\mathbf{C}_u = \mathbf{C}_u(d) = \mathbf{C}_u(E, d)$ denote the bounded real valued d-uniformly continuous functions on E. Then $\mathbf{C}_u$ consists precisely of the restrictions to E of the functions in $\mathbf{C}(\hat{E})$ - the (necessarily bounded) real valued continuous functions on $\hat{E}$. Since $\hat{E}$ is a compact metric space, it is clear that in the uniform norm $\mathbf{C}_u$ is a *separable* closed subalgebra and sublattice of $\mathbf{C}_b = \mathbf{C}_b(E)$ which separates the points of E. Of course, $\mathbf{C}_b$ itself is not separable in general. Following our general scheme of notation $\underline{\underline{E}}$

(resp. $\hat{E}$) denotes the σ-algebra of Borel subsets of E (resp. $\hat{E}$), while $\underline{\underline{E}}^*$ (resp. $\hat{\underline{\underline{E}}}^*$) denotes the σ-algebra of universally measurable subsets of E (resp. $\hat{E}$). Since E is a U-space $E \in \hat{\underline{\underline{E}}}^*$.

In discussing the relationship among these σ-algebras we shall need an important measure theoretic construction that will also be used several times in the sequel. Therefore we shall formulate it explicitly in the next several paragraphs.

Let $(F, \underline{\underline{F}})$ be an arbitrary measurable space and E be a subset of F. It is not assumed that $E \in \underline{\underline{F}}$. The trace of $\underline{\underline{F}}$ on E, denoted by $\underline{\underline{F}}|_E$, is the σ-algebra on E which consists of those subsets $B \subset E$ such that there exists $A \in \underline{\underline{F}}$ with $B = A \cap E$. We leave it to the reader to verify that $\underline{\underline{F}}|_E$ is indeed a σ-algebra on E. Of course, the representation of $B \in \underline{\underline{F}}|_E$ as $B = A \cap E$ with $A \in \underline{\underline{F}}$ is not unique. If $E \in \underline{\underline{F}}$, then $B \in \underline{\underline{F}}|_E$ if and only if $B \subset E$ and $B \in \underline{\underline{F}}$. If f is a numerical function on F, then $f|_E$ denotes the restriction of f to E. It is clear that a numerical function g on E is $\underline{\underline{F}}|_E$ measurable if and only if there exists $f \in \underline{\underline{F}}$ with $g = f|_E$.

Let $(F, \underline{\underline{F}})$ and E be as in the preceding paragraph. For typographical convenience set $\underline{\underline{B}} = \underline{\underline{F}}|_E$. Let λ be a finite measure on $(F, \underline{\underline{F}})$. For each $B \in \underline{\underline{B}}$ define

$$(8.2) \qquad \mu(B) = \inf\{\lambda(A) \colon A \supset B,\ A \in \underline{\underline{F}}\}.$$

We shall show that μ is a measure on $(E, \underline{\underline{B}})$. It is called the trace of λ on E. We need a bit more notation before coming to the proof. Let $\underline{\underline{\Gamma}}_E$ be the collection of all sets in $\underline{\underline{F}}$ which contain E. Clearly there exist sets in $\underline{\underline{\Gamma}}_E$ of minimal λ measure. (Take a decreasing sequence $(A_n) \subset \underline{\underline{\Gamma}}_E$ with $\inf \lambda(A_n) = \inf\{\lambda(A) \colon A \in \underline{\underline{\Gamma}}_E\}$. Then $A = \cap A_n \in \underline{\underline{\Gamma}}_E$ and has minimal λ measure.) The following lemma contains the facts that we shall need about the set function μ defined in (8.2).

(8.3) LEMMA. *μ is a measure on $(E, \underline{\underline{B}})$. Moreover if A_0 is any set in $\underline{\underline{\Gamma}}_E$ of minimal λ measure and $B \in \underline{\underline{B}}$ has the representation $B = A \cap E$ with $A \in \underline{\underline{F}}$, then $\mu(B) = \lambda(A \cap A_0)$. μ is called the trace of λ on E.*

PROOF. Let A_0 be any fixed set in $\underline{\underline{\Gamma}}_E$ of minimal λ measure. If $B \in \underline{\underline{B}}$ and $B = A \cap E$ with $A \in \underline{\underline{F}}$, define $\nu(B) = \lambda(A \cap A_0)$. We claim that $\nu(B)$ is well defined; that is, it does not depend on the choice of A in the representation

$B = A \cap E$. To see this support $B = A_1 \cap E = A_2 \cap E$ with $A_1, A_2 \in \underline{\underline{F}}$. Replacing A_1 by $A_1 \cap A_2$ we may assume that $A_1 \subset A_2$ without loss of generality. Then $(A_2 - A_1) \cap E$ is empty which implies $(A_2 - A_1)^c \supset E$. (Here $A^c = F - A$.) But then $\lambda(A_0) = \lambda[A_0 \cap (A_2 - A_1)^c]$, or equivalently, $\lambda[A_0 \cap (A_2 - A_1)] = 0$. Hence $\lambda(A_0 \cap A_2) = \lambda(A_0 \cap A_1)$, showing that ν is well defined. It is now easy to check that ν is a measure on $\underline{\underline{B}}$ and that for each $B \in \underline{\underline{B}}$, $\nu(B) = \mu(B)$ where μ is defined in (8.2). Since A_0 was an arbitrary element of $\underline{\underline{\Gamma}}_E$ of minimal λ measure the proof of (8.3) is complete.

We return now to our discussion of U-spaces. Using the notation developed in the paragraph following (8.1), the next proposition collects the basic facts about a U-space E contained in a compact metric space $\hat{E}$ as a universally measurable subspace that we shall need.

(8.4) PROPOSITION. (i) *$\underline{\underline{E}}$ is the trace of $\hat{\underline{\underline{E}}}$ on E.*

(ii) $\sigma(\underset{\sim}{C}_u) = \underline{\underline{E}}$.

(iii) *$\underline{\underline{E}}^*$ is the trace of $\hat{\underline{\underline{E}}}^*$ on E.*

PROOF. By the definition of the subspace topology, G is open in E if and only if there exists $\hat{G}$ open in $\hat{E}$ with $G = \hat{G} \cap E$. Statement (i) is an easy consequence of this. If $f \in \underset{\sim}{C}_u$, then $f = \hat{f}|_E$ with $\hat{f} \in \underset{\sim}{C}(\hat{E})$, and so by (i) f is $\underline{\underline{E}}$ measurable. Therefore $\sigma(\underset{\sim}{C}_u) \subset \underline{\underline{E}}$. Let $\underset{\sim}{H} = \{\hat{f} \in b\hat{\underline{\underline{E}}}: \hat{f}|_E \in \sigma(\underset{\sim}{C}_u)\}$. Then $\underset{\sim}{C}(\hat{E}) \subset \underset{\sim}{H}$ and it is immediate from the monotone class theorem that $\underset{\sim}{H} = b\hat{\underline{\underline{E}}}$. Therefore the trace of $\hat{\underline{\underline{E}}}$ on E is contained in $\sigma(\underset{\sim}{C}_u)$ and so by (i), $\underline{\underline{E}} \subset \sigma(\underset{\sim}{C}_u)$ proving (ii). Finally (iii) is an immediate consequence of (i) and the following lemma which is of interest in its own right.

(8.5) LEMMA. *Let $(F, \underline{\underline{F}})$ be a measurable space and let $E \subset F$. Let $\underline{\underline{B}}$ be the trace of $\underline{\underline{F}}$ on E and let $\underline{\underline{T}}^*$ be the trace of $\underline{\underline{F}}^*$ on E. Then $\underline{\underline{T}}^* \subset \underline{\underline{B}}^*$, and if $E \in \underline{\underline{F}}^*$, then $\underline{\underline{T}}^* = \underline{\underline{B}}^*$.*

PROOF. Let $A \in \underline{\underline{F}}^*$ and let $B = A \cap E$. Let μ be a finite measure on $(E, \underline{\underline{B}})$. Define ν on $(F, \underline{\underline{F}})$ by $\nu(C) = \mu(C \cap E)$ for all $C \in \underline{\underline{F}}$. Since $\underline{\underline{B}}$ is the trace of $\underline{\underline{F}}$ on E this makes sense and ν is a finite measure on $(F, \underline{\underline{F}})$. Thus there exist $A_1, A_2 \in \underline{\underline{F}}$ with $A_1 \subset A \subset A_2$ and $\nu(A_2) = \nu(A_1)$. Let $B_j = A_j \cap E \in \underline{\underline{B}}$ for $j = 1, 2$. Then $B_1 \subset B \subset B_2$ and $\mu(B_j) = \nu(A_j)$. Hence $B \in \underline{\underline{B}}^*$, and so $\underline{\underline{T}}^* \subset \underline{\underline{B}}^*$.

Now assume $E \in \mathcal{F}^*$. Let $B \in \mathcal{B}^*$ and let λ be a measure on $(F, \mathcal{F})$. Since $E \in \mathcal{F}^*$ there exist $E_1, E_2 \in \mathcal{F}$ with $E_1 \subset E \subset E_2$ and $\lambda(E_2) = \lambda(E_1)$. Clearly E_2 is a set in $\mathcal{F}$ containing E of minimal λ measure. Let μ be the trace of λ on $(E, \mathcal{B})$ constructed in (8.3). Then there exist $B_1, B_2 \in \mathcal{B}$ with $B_1 \subset B \subset B_2$ and $\mu(B_2) = \mu(B_1)$. Since $\mathcal{B}$ is the trace of $\mathcal{F}$ on E, there exist $A_j \in \mathcal{F}$ with $B_j = A_j \cap E$ for $j = 1, 2$. Let $C_j = A_j \cap E_j$, $j = 1, 2$. Then $C_1, C_2 \in \mathcal{F}$, $C_1 \subset B \subset C_2$, and $\lambda(C_1) = \lambda(A_1 \cap E_1) = \lambda(A_1 \cap E_2) = \mu(B_1)$ while $\lambda(C_2) = \lambda(A_2 \cap E_2) = \mu(B_2)$. Hence $B \in \mathcal{F}^*$, completing the proof of (8.5).

REMARK. The fact that $E \in \hat{\mathcal{E}}^*$ was used only in the proof of (8.4-iii). Consequently the other statements in (8.4) are valid for an *arbitrary* subspace E of $\hat{E}$.

The following fact will be used frequently. It is an immediate consequence of (8.4-ii) and D-IV-T18 (or I-T20 of [8]).

(8.6) LEMMA. *Let $\mathbf{H}$ be a vector space of bounded functions on E with $1 \in \mathbf{H}$. Suppose that $\mathbf{H}$ is closed under uniform convergence and that if $0 \le f_n \uparrow f$ with $(f_n) \subset \mathbf{H}$ and f bounded, then $f \in \mathbf{H}$. If $\mathbf{H}$ contains $\mathbf{C}_u$, then $\mathbf{H}$ contains all bounded $\mathcal{E}$ measurable functions on E.*

It is evident from (8.6) that a finite measure on $(E, \mathcal{E})$ is determined by its values on $\mathbf{C}_u$.

We next state Lusin's theorem (sometimes called Kuratowski's theorem) for ease of future reference. We refer the reader to [13] for a proof where it appears as Corollary 3.3 on page 22. Also see [2].

(8.7) THEOREM. *Let E be a complete separable metric space, B a Borel subset of E, and F a separable metric space. Let φ be an injection of B into F that is measurable; that is, for each Borel subset A of F, $\varphi^{-1}(A)$ is Borel in B, or equivalently in E. Then $\varphi(B)$ is Borel in F.*

We now recall the notion of a projective system. See BG-I-(2.9). Let T be an arbitrary index set and $(E, \mathcal{E})$ a measurable space. Let $\Phi(T)$ denote the class of all nonempty finite subsets of T. If $J \in \Phi(T)$, let $(E^J, \mathcal{E}^J)$ denote the usual finite product measurable space. If $J, K \in \Phi(T)$ and $J \subset K$, let π^K_J be the canonical projection of E^K on E^J. In particular π^K_J is $\mathcal{E}^K | \mathcal{E}^J$ measurable. For each $J \in \Phi(T)$, let μ_J be a probability measure on $(E^J, \mathcal{E}^J)$. Then the collection $\{\mu_J: J \in \Phi(T)\}$ is called a <u>projective system</u> over $(E, \underline{\underline{E}})$ provided

(8.8) $$\pi^K_J(\mu_K) = \mu_J; \ J, K \in \Phi(T), \ J \subset K .$$

(Recall that if f is a measurable map from a measurable space $(A, \underline{\underline{A}})$ to a measurable space $(B, \underline{\underline{B}})$ and μ is a measure on $(A, \underline{\underline{A}})$, then the image of μ under f, denoted by $f(\mu)$, is the measure ν on $(B, \underline{\underline{B}})$ defined by $\nu(C) = \mu[f^{-1}(C)]$ for all $C \in \underline{\underline{B}}$.)

Suppose $(\Omega, \underline{\underline{F}}, P)$ is a probability space and that for each $t \in T$, $X_t: \Omega \to E$ is $\underline{\underline{F}}|\underline{\underline{E}}$ measurable, that is, each X_t is an $(E, \underline{\underline{E}})$ random variable. If $J \in \Phi(T)$, $J = (t_1, \ldots, t_n)$ define $X_J: \Omega \to E^J$ by $X_J(\omega) = (X_{t_1}(\omega), \ldots, X_{t_n}(\omega)) \in E^J$. Then X_J is $\underline{\underline{F}}|\underline{\underline{E}}^J$ measurable and if we define $P_J = X_J(P)$, the collection $\{P_J: J \in \Phi(T)\}$ is plainly a projective system over $(E, \underline{\underline{E}})$ called <u>the system of finite dimensional distributions of the process</u> X.

We need one less familiar definition. Let $(E, \underline{\underline{E}}, \mu)$ be a probability space and suppose that $f: \Omega \to E$ where Ω is an abstract set. Then f is <u>almost surjective</u> if $f(\Omega)$ has μ outer measure one in E.

We are now ready to formulate our next result.

(8.9) LEMMA. <u>Let</u> T <u>be an arbitrary set,</u> $(E, \underline{\underline{E}})$ <u>a measurable space, and</u> $\{\mu_J: J \in \Phi(T)\}$ <u>a given projective system over</u> $(E, \underline{\underline{E}})$. <u>Let</u> Ω <u>be an abstract set and suppose that for each</u> $J \in \Phi(T)$ <u>we are given a map</u> $X_J: \Omega \to E^J$ <u>satisfying</u> $X_J = \pi^K_J \circ X_K$ <u>whenever</u> $K, J \in \Phi(T)$ <u>with</u> $J \subset K$. <u>Let</u>

(8.10) $$\underline{\underline{A}} = \{\Lambda \subset \Omega: \Lambda = X_J^{-1}(B_J), \ B_J \in \underline{\underline{E}}^J, \ J \in \Phi(T)\} .$$

<u>Then</u> $\underline{\underline{A}}$ <u>is an algebra on</u> Ω. <u>If, in addition, each</u> X_J <u>is almost surjective, then there exists a unique finitely additive probability</u> Q <u>on</u> $(\Omega, \underline{\underline{A}})$ <u>such that</u> $\mu_J = X_J(Q)$ <u>for each</u> $J \in \Phi(T)$.

PROOF. This is essentially the first step in the proof of the Kolmogorov existence theorem. We leave it to the reader to verify that $\underline{\underline{A}}$ is an algebra and content ourselves with constructing Q. If $\Lambda \in \underline{\underline{A}}$, then $\Lambda = X_J^{-1}(B_J)$ for some $J \in \Phi(T)$ and $B_J \in \underline{\underline{E}}^J$. Let $Q(\Lambda) = \mu_J(B_J)$. We must check that this is well defined; that is, if in addition $\Lambda = X_K^{-1}(B_K)$ with $B_K \in \underline{\underline{E}}^K$, then $\mu_J(B_J) = \mu_K(B_K)$. Let $L = K \cup J$, $B^J_L = (\pi^L_J)^{-1}(B_J)$, and $B^K_L = (\pi^L_K)^{-1}(B_K)$. Then B^J_L and B^K_L

are in $\underline{\underline{E}}^L$, and

$$X_L^{-1}(B_L^J) = X_L^{-1}[(\pi_J^L)^{-1}(B_J)] = (\pi_J^L \circ X_L)^{-1}(B_J) = X_J^{-1}(B_J) = \Lambda .$$

Therefore $X_L(\Lambda) \subset B_L^J$. Let $F^L = X_L(\Omega)$. Then if $x \in B_L^J \cap F^L$, there exists an ω such that $X_L(\omega) = x \in B_L^J$ and so $\omega \in X_L^{-1}(B_L^J) = \Lambda$. As a result $X_L(\Lambda) \cap F^L = B_L^J \cap F^L$. Similarly, $X_L(\Lambda) \cap F^L = B_L^K \cap F^L$. Let $\tilde{\mu}_L$ be the trace of μ_L on F^L as defined in (8.3). Since F^L has outer μ_L measure one, E^L is a set containing F^L of minimal μ_L measure, and so

$$\tilde{\mu}_L(B_L^J \cap F^L) = \mu_L(B_L^J \cap E^L) = \mu_L(B_L^J) = \mu_L[(\pi_J^L)^{-1}(B_J)] = \mu_J(B_J) .$$

Similarly $\tilde{\mu}_L(B_L^K \cap F^L) = \mu_K(B_K)$, and since $B_L^J \cap F^L = B_L^K \cap F^L$ it follows that $\mu_J(B_J) = \mu_K(B_K)$. Thus Q is well defined and we leave the reader the straightforward task of verifying that Q is a finitely additive probability on $\underline{\underline{A}}$ with $\mu_J = X_J(Q)$ for each $J \in \Phi(T)$.

We come now to the key lemma. Let $(\Omega, \underline{\underline{F}}, (X_t)_{t \in T}, P)$ be a stochastic process with state space $(E, \underline{\underline{E}})$. Again the index set T is an arbitrary set and $(E, \underline{\underline{E}})$ is an abstract measurable space. We assume that $\underline{\underline{F}} = \sigma(X_t;\ t \in T)$. Let $X_J\colon \Omega \to E^J$ be defined as before, $X_J(\omega) = (X_{t_1}(\omega), \ldots, X_{t_n}(\omega))$ if $J = (t_1, \ldots, t_n) \in \Phi(T)$, and let $(\mu_J : J \in \Phi(T))$ be the projective system of finite dimensional distributions of X. Let Ω_0 be a subset of Ω and for each $J \in \Phi(T)$ let X_J^0 be the restriction of X_J to Ω_0. Clearly if $J, K \in \Phi(T)$ and $J \subset K$, $X_J^0 = \pi_J^K \circ X_K^0$. Let $\underline{\underline{A}}_0$ be the algebra on Ω_0 defined as in (8.10) but using the maps X_J^0.

(8.11) LEMMA. Using the above notation, assume that each X_J^0 is almost surjective and let Q be the finitely additive probability on $\underline{\underline{A}}_0$ defined in (8.9). Then Q is countably additive if and only if Ω_0 has P outer measure one. In this case Q is the trace of P on Ω_0 restricted to $\underline{\underline{A}}_0$.

PROOF. Let $\underline{\underline{A}}$ be the algebra on Ω defined in (8.10). Then $\underline{\underline{A}} \subset \underline{\underline{F}}$, $\sigma(\underline{\underline{A}}) = \underline{\underline{F}}$, and $\underline{\underline{A}}_0$ is the trace of $\underline{\underline{A}}$ on Ω_0. If $\Lambda \in \underline{\underline{A}}$, then $\Lambda = X_J^{-1}(B_J)$ for some $J \in \Phi(T)$ and $B_J \in \underline{\underline{E}}^J$. If $\Lambda_0 = \Lambda \cap \Omega_0$, then $\Lambda_0 = (X_J^0)^{-1}(B_J)$, and so $P(\Lambda) = \mu_J(B_J) = Q(\Lambda_0)$. In other words $Q(\Lambda \cap \Omega_0) = P(\Lambda)$ for all Λ in $\underline{\underline{A}}$.

Now suppose that Ω_0 has P outer measure one. Then Ω is a set of minimal P measure containing Ω_0 and so if P^0 denotes the trace of P on Ω_0 one has $P^0(\Gamma \cap \Omega_0) = P(\Gamma)$ for all $\Gamma \in \underline{\underline{F}}$. Consequently Q and P^0 agree on $\underline{\underline{A}}_0$, and since P^0 is countably additive so is Q. Conversely if Q is countably additive on $\underline{\underline{A}}_0$, then Q extends as a measure to $\sigma(\underline{\underline{A}}_0)$. If $(\Lambda_j) \subset \underline{\underline{A}}$ and $\Omega_0 \subset \cup \Lambda_j$, then $\Omega_0 = \cup(\Lambda_j \cap \Omega_0)$ and so

$$1 = Q(\Omega_0) \le \sum Q(\Lambda_j \cap \Omega_0) = \sum P(\Lambda_j) .$$

Therefore Ω_0 has P outer measure one. If $\Lambda_0 \in \underline{\underline{A}}_0$ then $\Lambda_0 = \Omega_0 \cap \Lambda$ with $\Lambda \in \underline{\underline{A}}$, and $Q(\Lambda_0) = P(\Lambda)$. Thus Q is the trace of P on Ω_0 restricted to $\underline{\underline{A}}_0$, completing the proof of (8.11).

9. RIGHT PROCESSES

In this section we introduce the class of processes that shall concern us in the remainder of these lectures. These are essentially the processes satisfying the "hypothèses droites" of Meyer except that we shall only assume that E is a U-space rather than Lusinien, and we shall drop the requirement that the excessive functions be nearly Borel.

We fix a U-space E and we shall regard E as a universally measurable subspace of a compact metric space $\hat{E}$. We fix a metric d on $\hat{E}$ compatible with the topology of $\hat{E}$ and we shall use the notation of the first part of Section 8. We remind the reader that $\underset{\sim}{C}_u = \underset{\sim}{C}_u(E, d)$ denotes the bounded real valued d-uniformly continuous functions on E, or equivalently the restrictions to E of the real valued continuous functions on $\hat{E}$.

We now fix a Markov semigroup $(P_t)_{t \geq 0}$ on $(E, \underline{\underline{E}}^*)$. The important point here is that we only assume $x \to P_t(x, B)$ is $\underline{\underline{E}}^*$ measurable even when B is in $\underline{\underline{E}}$. Let Ω be the space of all right continuous maps $\omega: \mathbb{R}^+ \to E$. No assumption about the existence of left limits is made. As usual let $X_t(\omega) = \omega(t)$ be the coordinate maps. Thus for each $t \in \mathbb{R}^+$, $X_t: \Omega \to E$. As is customary, for typographical convenience, we shall sometimes write $X(t)$ for X_t. Let $\underline{\underline{F}}^0$, resp. $\underline{\underline{F}}_t^0$, be the σ-algebra on Ω generated by $(X_s: s \geq 0)$, resp. $(X_s: s \leq t)$, when each X_s is regarded as a map from Ω to $(E, \underline{\underline{E}})$, and let $\underline{\underline{F}}^{0*}$ and $\underline{\underline{F}}_t^{0*}$ be defined similarly when each X_s is regarded as a map from Ω to $(E, \underline{\underline{E}}^*)$. That is $\underline{\underline{F}}^0$, resp. $\underline{\underline{F}}^{0*}$, is generated by all sets of the form $X_s^{-1}(A)$ with $s \geq 0$ and $A \in \underline{\underline{E}}$, resp. $A \in \underline{\underline{E}}^*$, and similarly for $\underline{\underline{F}}_t^0$ and $\underline{\underline{F}}_t^{0*}$.

We are now ready to state our first hypothesis on the semigroup (P_t). As is customary a measure on $\underline{\underline{E}}$ is automatically (and uniquely) extended to $\underline{\underline{E}}^*$. Thus there is no difference between giving a measure on $\underline{\underline{E}}$ or on $\underline{\underline{E}}^*$.

HD1. (HYPOTHÈSES DROITES 1.) <u>For each probability</u> μ <u>on</u> $(E, \underline{\underline{E}})$ <u>there exists a probability</u> P^μ <u>on</u> $(\Omega, \underline{\underline{F}}^{0*})$ <u>such that</u> $(X_t, \underline{\underline{F}}_t^{0*}, P^\mu)$ <u>is a Markov</u>

process with initial measure μ and transition function P_t; that is,

(9.1) $$P^{\mu}(X_0 \in A) = \mu(A) \qquad \text{for } A \in \underline{\underline{E}}^*.$$

(9.2) $$E^{\mu}(f(X_{t+s}) | \underline{\underline{F}}_s^{0*}) = P_t f(X_s) \qquad \text{for } t, s \geq 0 \text{ and } f \in b\underline{\underline{E}}^*.$$

REMARKS. Since $P_t f \in b\underline{\underline{E}}^*$ when $f \in b\underline{\underline{E}}^*$, the right side of (9.2) is $\underline{\underline{F}}_s^{0*}$ measurable. In general, even for $f \in \underset{\sim}{C}_u$, $P_t f(X_s)$ is not $\underline{\underline{F}}_s^0$ measurable. When $\mu = \epsilon_x$ we write P^x and E^x for P^{μ} and E^{μ}. Introducing the shift operators $(\theta_t)_{t \geq 0}$ on Ω by $\theta_t \omega(s) = \omega(t+s)$, (9.2) extends in the usual manner to

(9.3) $$E^{\mu}(Z \circ \theta_s | \underline{\underline{F}}_s^{0*}) = E^{X(s)}(Z)$$

for all $Z \in b\underline{\underline{F}}^{0*}$. Here $E^{X(s)}(Z)$ is the evaluation of the function $x \to E^x(Z)$ at the point $X_s = X(s)$. Part of the verification of (9.3) involves checking that this function is $\underline{\underline{E}}^*$ measurable so that the right side of (9.3) is $\underline{\underline{F}}_s^{0*}$ measurable. If $\mu = \epsilon_x$, combining (9.1) and (9.2) gives

$$f(x) = E^x[f(X_0)] = E^x[P_0 f(X_0)] = P_0 f(x),$$

and so $P_0 = I$.

Since $P_t f(x) = E^x[f(X_t)]$ and $t \to X_t$ is right continuous, we see that for $f \in \underset{\sim}{C}_u$, $t \to P_t f(x)$ is right continuous. It follows from this and (8.6) that $(t, x) \to P_t f(x)$ is $\underline{\underline{R}}^+ \otimes \underline{\underline{E}}^*$ measurable for $f \in b\underline{\underline{E}}$. If $f \in b\underline{\underline{E}}^*$, then $(t, x) \to P_t f(x)$ is $(\underline{\underline{R}}^+ \otimes \underline{\underline{E}}^*)^{\lambda, \mu}$ measurable for all finite measures λ on $\underline{\underline{R}}^+$ and μ on $\underline{\underline{E}}^*$, where the σ-algebra in question is the completion of $\underline{\underline{R}}^+ \otimes \underline{\underline{E}}^*$ with respect to the product measure $\lambda \otimes \mu$. To see this define $\nu(A) = \int_0^\infty \lambda(dt) \int \mu(dx) P_t(x, A)$ for $A \in \underline{\underline{E}}$. Since $f \in b\underline{\underline{E}}^*$ there exist $f_1, f_2 \in b\underline{\underline{E}}$ with $f_1 \leq f \leq f_2$ and $\nu(f_2) = \nu(f_1)$. Therefore $P_t f_1(x) \leq P_t f(x) \leq P_t f_2(x)$ for all (t, x), the extremes are $\underline{\underline{R}}^+ \otimes \underline{\underline{E}}^*$ measurable, and by the definition of ν they agree almost everywhere relative to $\lambda \otimes \mu$. This proves that $(t, x) \to P_t f(x)$ is $(\underline{\underline{R}}^+ \otimes \underline{\underline{E}}^*)^{\lambda, \mu}$ measurable. Exactly the same argument shows that $t \to P_t f(x)$ is Lebesgue measurable for each $f \in b\underline{\underline{E}}^*$ and $x \in E$.

We now introduce the definitive σ-algebras on Ω. For each μ let $\underline{\underline{F}}^{\mu}$ be the completion of $\underline{\underline{F}}^{0*}$ with respect to P^{μ}, and $\underline{\underline{F}}_t^{\mu}$ be the σ-algebra generated by $\mathcal{F}_t^{0*}$ and all sets in $\mathcal{F}^\mu$ of P^μ measure zero. Let $\mathcal{F} = \cap\, \mathcal{F}^\mu$ and $\mathcal{F}_t = \cap\, \mathcal{F}_t^\mu$ where in both cases the intersection is over all probabilities μ on $(E, \mathcal{E})$. Actually $\mathcal{F}^\mu$ and $\mathcal{F}_t^\mu$ are unchanged if one replaces $\mathcal{F}^{0*}$ and $\mathcal{F}_t^{0*}$ by $\mathcal{F}^0$ and $\mathcal{F}_t^0$ respectively in the above definitions. To see this let $\mathcal{G}^\mu$ denote the P^μ completion of $\mathcal{F}^0$. Since $\mathcal{F}^0 \subset \mathcal{F}^{0*}$, $\mathcal{G}^\mu \subset \mathcal{F}^\mu$. For the reverse inclusion let $f \in b\mathcal{E}^*$ and $t \ge 0$. If $\nu = \mu P_t$, then there exist $f_1, f_2 \in b\mathcal{E}$ with $f_1 \le f \le f_2$ and $\nu(f_1) = \nu(f_2)$. Now $f_1 \circ X_t \le f \circ X_t \le f_2 \circ X_t$ and

$$E^\mu[f_2 \circ X_t - f_1 \circ X_t] = \int \mu(dx)\, P_t(f_2 - f_1)(x) = \nu(f_2 - f_1) = 0 \ ,$$

and so $f(X_t)$ is $\mathcal{G}^\mu$ measurable. Thus $\mathcal{F}^{0*} \subset \mathcal{G}^\mu$ and so $\mathcal{F}^\mu \subset \mathcal{G}^\mu$. Hence $\mathcal{F}^\mu = \mathcal{G}^\mu$. A similar argument shows that $\mathcal{F}_t^\mu$ is the σ-algebra generated by $\mathcal{F}_t^0$ and all P^μ null sets in $\mathcal{F}^\mu$. Therefore when one passes to the definitive σ-algebras $\mathcal{F}^\mu$ and $\mathcal{F}_t^\mu$ the need for $\mathcal{F}_t^{0*}$ and $\mathcal{F}^{0*}$ disappears.

As in Section I. 5 of BG one checks that $x \to E^x(Z)$ is $\mathcal{E}^*$ measurable for $Z \in b\mathcal{F}$, and that $E^\mu(Z \circ \theta_s \mid \mathcal{F}_s^\mu) = E^{X(s)}(Z)$ for such Z. Taking expectations in this last equality with $s = 0$ yields $E^\mu(Z) = \int E^x(Z)\, \mu(dx)$; a fact that we shall often use without special mention. If $f \in b\mathcal{E}^*$, then $t \to P_t f(x)$ is Lebesgue measurable and so one can define

$$U^\alpha f(x) = \int_0^\infty e^{-\alpha t}\, P_t f(x)\, dt \ .$$

There is enough joint measurability to check that $(U^\alpha)_{\alpha > 0}$ is a Markov resolvent on $(E, \mathcal{E}^*)$, which uniquely determines $(P_t)_{t \ge 0}$ by (8.6), since $t \to P_t f(x)$ is right continuous for $f \in \underset{\sim}{C}_u$. Using the right continuity of $t \to f \circ X_t(\omega)$ for $f \in \underset{\sim}{C}_u$ and (8.6) one checks that if $f \in b\mathcal{E}^*$, then $(t, \omega) \to f \circ X_t(\omega)$ is measurable with respect to the $\lambda \otimes P^\mu$ completion of $\mathbb{R}^+ \otimes \mathcal{F}^{0*}$ for all finite measures λ and μ on $\mathbb{R}^+$ and $\mathcal{E}$ respectively. Consequently one may apply Fubini's theorem to see that

$$U^\alpha f(x) = E^x \int_0^\infty e^{-\alpha t} f(X_t)\, dt \ .$$

We are now in position to state the second hypothesis that we shall impose on the semigroup (P_t) .

HD2. HYPOTHÈSES DROITES 2.) Let f be α-excessive (for the resolvent (U^α) on $(E, \underline{\underline{E}}^*)$). Then for each probability μ on $(E, \underline{\underline{E}})$, $t \to f(X_t)$ is P^μ almost surely right continuous on $\mathbb{R}^+$.

A nonnegative function f on E is nearly Borel for X, or (P_t), if for each probability μ on $(E, \underline{\underline{E}})$ there exist $f_1, f_2 \in \underline{\underline{E}}$ with $f_1 \le f \le f_2$ such that the processes $(f_1 \circ X_t)$ and $(f_2 \circ X_t)$ are P^μ indistinguishable. Since $P_0 = I$ this implies that $f \in \underline{\underline{E}}^*$ and so it is not necessary to assume this explicitly as in (5.7).

(9.4) THEOREM. Assume that HD1 holds.

(i) Let μ be a probability on $(E, \underline{\underline{E}})$. Then HD2 implies that X is strong Markov with respect to P^μ and that $(\underline{\underline{F}}_t^\mu)$ is right continuous. Conversely if (P_t) sends $b\underline{\underline{E}}$ into $b\underline{\underline{E}}$ and if X is strong Markov, then HD2 holds and each α-excessive function is nearly Borel.

(ii) Let X be strong Markov. Then HD2 is equivalent to the statement that if f is α-excessive, then for each μ on $(E, \underline{\underline{E}})$ the process $f \circ X = (f \circ X_t)$ is P^μ indistinguishable from a well measurable process over $(\Omega, \underline{\underline{F}}^\mu, \underline{\underline{F}}_t^\mu, P^\mu)$.

PROOF. That HD2 implies that X is strong Markov relative to P^μ and that $(\underline{\underline{F}}_t^\mu)$ is right continuous is proved exactly as in BG-I-(8.11) and BG-I-(8.12) to which we refer the reader. In fact for this argument one only needs that $t \to U^\alpha g(X_t)$ is P^μ almost surely right continuous for $g \in \underset{\sim}{C}_u^+$ and $\alpha > 0$. (In the proof of BG-I-(8.11) take $\underset{\sim}{L} = \underset{\sim}{C}_u$, and replace the last two sentences of the argument with an appeal to (8.6).)

Next suppose that (P_t) sends $b\underline{\underline{E}}$ into $b\underline{\underline{E}}$, that is, each P_t is a (Markov) kernel on $(E, \underline{\underline{E}})$, and that X is strong Markov with respect to P^μ. By this we mean that if T is an $(\underline{\underline{F}}_{t+}^0)$ stopping time, then

$$(9.5) \qquad E^\mu[f(X_{t+T})\, 1_{\{T<\infty\}} \mid \underline{\underline{F}}_{T+}^0] = P_t f(X_T)\, 1_{\{T<\infty\}}$$

for each $f \in \underset{\sim}{C}_u^+$ and $t \ge 0$. Note that since $P_t f$ is Borel the right side of (9.5) is $\underline{\underline{F}}_{T+}^0$ measurable. Obviously (9.5) extends to all $f \in b\underline{\underline{E}}$, and according to BG-I-(8.12) it follows from (9.5) that $(\underline{\underline{F}}_t^\mu)$ is right continuous. Also (9.5) extends to $(\underline{\underline{F}}_t^\mu)$ stopping times T if we replace $\underline{\underline{F}}_{T+}^0$ by $\underline{\underline{F}}_T^\mu$ in its statement. See BG-I-(7.3). Now suppose $f = U^\alpha g$ with $g \in b\underline{\underline{E}}^+$ and $\alpha > 0$. Let (T_n) be a decreasing sequence of $(\underline{\underline{F}}_t^\mu)$ stopping times with limit T. Then using the strong Markov property and the bounded convergence theorem

$$E^{\mu}[e^{-\alpha T_n} U^{\alpha} g(X_{T_n})] = E^{\mu} \int_{T_n}^{\infty} e^{-\alpha t} g(X_t) dt$$

$$\to E^{\mu} \int_{T}^{\infty} e^{-\alpha t} g(X_t) dt = E^{\mu}[e^{-\alpha T} U^{\alpha} g(X_T)] .$$

Since $U^{\alpha} g \in b\underline{E}^{+}$, the process $(e^{-\alpha t} U^{\alpha} g \circ X_t)$ is well measurable relative to $(\Omega, \underline{F}^{\mu}, \underline{F}_t^{\mu}, P^{\mu})$ and so by D-IV-T28, $t \to e^{-\alpha t} U^{\alpha} g \circ X_t$ is P^{μ} almost surely right continuous, and because it is a bounded supermartingale it also has left limits almost surely. Now one can repeat the last part of the proof of Theorem 5.8 to show that HD2 holds and that the α-excessive functions are nearly Borel. This establishes (9.4-i).

If f is α-excessive, then $f \circ X_t$ is $\underline{F}_t^{0*}$ measurable. Therefore if HD2 holds, then for each μ, $f \circ X$ is P^{μ} indistinguishable from an $(\underline{F}_t^{\mu})$ adapted right continuous process, proving one of the implications in (9.4-ii). The statement that X is strong Markov in this case is that (9.5) holds for all $(\underline{F}_{t+}^{0*})$ stopping times T with $\underline{F}_{T+}^{0}$ replaced by $\underline{F}_{T+}^{0*}$. If $U^{\alpha} g \circ X_t$ is <u>assumed</u> to be (P^{μ} indistinguishable from) a well measurable process over $(\Omega, \underline{F}^{\mu}, \underline{F}_t^{\mu}, P^{\mu})$ then exactly the same argument shows that HD2 holds. This completes the proof of (9.4).

(9.6) REMARKS. Under HD1 and HD2 if f is α-excessive, then $t \to f \circ X_t$ has left limits on $(0, \infty)$ almost surely. If f is bounded this is the case because $(e^{-\alpha t} f \circ X_t)$ is a bounded almost surely right continuous supermartingale, and a passage to the limit using (4.1) gives the general case. We emphasize that under HD1 and HD2 the system $(\Omega, \underline{F}^{\mu}, \underline{F}_t^{\mu}, P^{\mu})$ satisfies the usual conditions of the general theory of processes (D-III-26) for each μ. It may be worthwhile to point out explicitly that the argument used in the proof of (5.8-iii) and (8.6) show that in HD2 it would suffice to assume that $t \to U^{\alpha} g(X_t)$ is P^{μ} almost surely right continuous on $\mathbb{R}^{+}$ for all probabilities μ on E, $g \in \underset{\sim}{C}_u$, and $\alpha > 0$. Finally the statement that two functions on $\mathbb{R}^{+} \times \Omega$ are <u>indistinguishable</u> means that they are P^{μ} indistinguishable for each μ.

(9.7) TERMINOLOGY. In the sequel we shall refer to a semigroup or a process satisfying HD1 and HD2 as a right semigroup or a right process. We shall say that a right semigroup or process is Borel if E is Lusinien and each P_t is a Markov kernel on $(E, \underline{\underline{E}})$; that is, $P_t : b\underline{\underline{E}} \to b\underline{\underline{E}}$ for each $t \geq 0$.

Right processes include most of the different classes of processes that have been studied within the framework of Markov processes on general state spaces. We now give a summary of the various types of processes encountered in probabilistic potential theory. In certain respects these definitions differ from those given in BG or in [9].

(9.8) A Hunt process is a Borel right process that is quasi-left-continuous; that is, if (T_n) is an increasing sequence of $(\underline{\underline{F}}_t)$ stopping times with limit T, then $X_{T_n} \to X_T$ almost surely on $\{T < \infty\}$. In BG it is assumed that E is compact. Note that the present state space E corresponds to E_Δ in BG.

(9.9) Let X be a right process and let Δ be a fixed point in E that acts as a trap; that is, if $\zeta = \inf\{t: X_t = \Delta\}$, then $X_t = \Delta$ for all $t \geq \zeta$. Then X is a standard process (relative to Δ) provided it is quasi-left-continuous on $[0, \zeta)$; that is, if (T_n) is an increasing sequence of $(\underline{\underline{F}}_t)$. stopping times with limit T, then $X_{T_n} \to X_T$ almost surely on $\{T < \zeta\}$. In [9] it is also assumed that E is Lusinien and that for each $\alpha \geq 0$ the α-excessive functions are nearly Borel. In BG it is assumed that (P_t) is Borel and that E is compact.

(9.10) A special standard process is a standard process such that for each μ the family $(\underline{\underline{F}}^\mu_t)$ is quasi-left-continuous (see D-III-D38).

(9.11) A Feller process is a Borel right process with compact state space E such that $U^\alpha : \underset{\sim}{C}(E) \to \underset{\sim}{C}(E)$ for each $\alpha > 0$ and $\alpha U^\alpha f \to f$ pointwise as $\alpha \to \infty$ for each $f \in \underset{\sim}{C}(E)$. Here $\underset{\sim}{C}(E)$ is the set of all real valued continuous functions on E. It is well known that a Feller process is quasi-left-continuous, and hence is a Hunt process. This is an immediate consequence of (5.15) since a Feller process is a Ray process with no branch points.

One has the following inclusions among these various classes of processes:

(Feller) $\subset$ (Hunt) $\subset$ (special standard) $\subset$ (standard) $\subset$ (right).

The only inclusion that is not clear is that a Hunt process is special standard. But this is essentially (6.7). See also BG-pages 171-172. These different types of processes were introduced at various stages during the development of the modern theory of Markov processes. In view of the theory to be developed in the following sections it seems to me that they are now mainly of historical interest. Probably the one subclass of right processes that should be singled out for special study is that for which the family $(\underline{\underline{F}}_t^{\mu})$ is quasi-left-continuous for each μ. Should these be called special right processes? For all practical purposes such a process is a Hunt process in the Ray topology to be introduced in the sequel. (See (13.3).)

Finally we point out that a Ray process Y on a compact state space E is not necessarily a right process since $P_0 \neq I$. However, if one restricts Y to the set of nonbranch points D, then it is a Borel right process. Since $Y_t \in D$ for all $t \geq 0$ this amounts to considering only initial measures μ that are carried by D. In what follows we shall see that the converse is true in the sense that if X is a right process with state space E, then by changing the topology on E one can essentially regard X as a Ray process restricted to its set of nonbranch points.

10. THE RAY KNIGHT COMPACTIFICATION

The assumptions are exactly the same as in Section 9. Thus X denotes a fixed right process with state space E, semigroup (P_t), and resolvent (U^α). All notation is the same as in Section 9. In particular E is a universally measurable subset of a compact metric space $(\hat{E}, d)$ and $\underset{\sim}{C}_u$ denotes the restrictions to E of the continuous functions on $\hat{E}$. Each P_t, $t \geq 0$ and αU^α, $\alpha > 0$ is a Markov kernel on $(E, \underline{\underline{E}}^*)$. In this section we shall introduce a new topology on E and it will turn out that, roughly speaking, X becomes a Ray process in this new topology.

We begin by constructing a new metric on E that makes the functions $U^\alpha f$ continuous for a sufficiently large class of f. We shall break up this construction into a number of steps.

(10.1) PROPOSITION. There exists a unique minimal convex cone $\underset{\sim}{R} = \underset{\sim}{R}(d)$ of bounded positive functions on E with the following properties:

(i) $\underset{\sim}{R} \subset b\underline{\underline{E}}^*_+$.

(ii) $U^\alpha \underset{\sim}{R} \subset \underset{\sim}{R}$ and $U^\alpha \underset{\sim}{C}_u^+ \subset \underset{\sim}{R}$ for each $\alpha > 0$.

(iii) $f, g \in \underset{\sim}{R} \Rightarrow f \wedge g \in \underset{\sim}{R}$.

(iv) $\underset{\sim}{R}$ is separable in the uniform topology.

REMARKS. $\underset{\sim}{R}$ is called the Ray cone of (P_t) or (U^α). Statement (iv) means that there is a countable subcollection $(f_n) \subset \underset{\sim}{R}$ such that given $f \in \underset{\sim}{R}$ and $\varepsilon > 0$ there exists a k with $\|f - f_k\| < \varepsilon$ where $\|g\| = \sup\{|g(x)| : x \in E\}$ for any numerical function g on E. Since $U^1 1 = 1$ it follows from (10.1-ii) that $\underset{\sim}{R}$ contains 1 and hence all positive constants. Of course, $\underset{\sim}{R} = \underset{\sim}{R}(d)$ depends on d or $\hat{E}$ through the condition $U^\alpha \underset{\sim}{C}_u^+ \subset \underset{\sim}{R}$.

PROOF. If $\underset{\sim}{H}$ is any convex cone contained in $b\underline{\underline{E}}^*_+$, define $\mathcal{U}(\underset{\sim}{H})$ and $\Lambda(\underset{\sim}{H})$ as follows:

$$\mathcal{U}(\underset{\sim}{H}) = \{U^{\alpha_1} f_1 + \cdots + U^{\alpha_n} f_n;\ \alpha_j > 0,\ f_j \in \underset{\sim}{H},\ 1 \le j \le n,\ n \ge 1\}$$

$$\Lambda(\underset{\sim}{H}) = \{f_1 \wedge \cdots \wedge f_n;\ f_j \in \underset{\sim}{H},\ 1 \le j \le n,\ n \ge 1\}\ .$$

It is evident that $\Lambda(\underset{\sim}{H})$ is closed under "$\wedge$", and, since $\underset{\sim}{H}$ is a convex cone, that $\mathcal{U}(\underset{\sim}{H})$ is a convex cone contained in $b\underset{=}{E}^*_+$. In addition if $\underset{\sim}{H}$ is separable for the uniform topology so is $\mathcal{U}(\underset{\sim}{H})$. In fact finite sums of the form $U^{\alpha_1} f_1 + \cdots + U^{\alpha_n} f_n$ as the α_j's range over the positive rationals and the f_j's over a countable dense subset of $\underset{\sim}{H}$ form a countable dense subset of $\mathcal{U}(\underset{\sim}{H})$. It is clear that $\Lambda(\underset{\sim}{H})$ is separable if $\underset{\sim}{H}$ is, and we claim that $\Lambda(\underset{\sim}{H})$ is a convex cone. If $f = f_1 \wedge \cdots \wedge f_n \in \Lambda(\underset{\sim}{H})$ and $\beta > 0$, then $\beta f = (\beta f_1) \wedge \cdots \wedge (\beta f_n) \in \Lambda(\underset{\sim}{H})$. Next suppose that $f, g, h \in \underset{\sim}{H}$. Then $f + (g \wedge h) = (f+g) \wedge (f+h) \in \Lambda(\underset{\sim}{H})$. If $f, g_1, \ldots, g_n, g_{n+1} \in \underset{\sim}{H}$ and $f + (g_1 \wedge \cdots \wedge g_n) \in \Lambda(\underset{\sim}{H})$, then

$$f + (g_1 \wedge \cdots \wedge g_n \wedge g_{n+1}) = [f + (g_1 \wedge \cdots \wedge g_n)] \wedge [f + g_{n+1}]$$

is in $\Lambda(\underset{\sim}{H})$. Consequently by induction $f+g \in \Lambda(\underset{\sim}{H})$ whenever $f \in \underset{\sim}{H}$ and $g \in \Lambda(\underset{\sim}{H})$. A similar induction on the first variable shows that $f+g \in \Lambda(\underset{\sim}{H})$ whenever $f, g \in \Lambda(\underset{\sim}{H})$. Therefore $\Lambda(\underset{\sim}{H})$ is a convex cone. Also note that if $\underset{\sim}{A}$ and $\underset{\sim}{B}$ are separable convex cones contained in $b\underset{=}{E}^*_+$, then

$$\underset{\sim}{A} + \underset{\sim}{B} = \{f + g:\ f \in \underset{\sim}{A},\ g \in \underset{\sim}{B}\}$$

is a separable convex cone contained in $b\underset{=}{E}^*_+$.

We now are ready to define $\underset{\sim}{R}$. Let $\underset{\sim}{R}^0 = \mathcal{U}(\underset{\sim}{C}^+_u)$, $\underset{\sim}{R}^1 = \Lambda(\underset{\sim}{R}^0 + \mathcal{U}(\underset{\sim}{R}^0))$, $\ldots$, $\underset{\sim}{R}^{n+1} = \Lambda(\underset{\sim}{R}^n + \mathcal{U}(\underset{\sim}{R}^n))$, $\ldots$, and $\underset{\sim}{R} = \cup \underset{\sim}{R}^n$. Note that $\underset{\sim}{R}^n \subset \underset{\sim}{R}^{n+1}$ for each $n \ge 0$. Since $\underset{\sim}{C}^+_u$ is a separable convex cone it is clear that $\underset{\sim}{R}^0$ also is a separable convex cone with $1 \in \underset{\sim}{R}^0$ and $\underset{\sim}{R}^0 \subset b\underset{=}{E}^*_+$. It follows by induction that each $\underset{\sim}{R}^n$ is a separable convex cone contained in $b\underset{=}{E}^*$ and hence so is $\underset{\sim}{R}$. Thus $\underset{\sim}{R}$ satisfies (i) and (iv) of (10.1). If $f, g \in \underset{\sim}{R}$, then there exists an $n \ge 1$ with $f, g \in \underset{\sim}{R}^n$, and so $f \wedge g \in \underset{\sim}{R}^n$. Thus (10.1-iii) holds. Now $U^\alpha(\underset{\sim}{C}^+_u) \subset \underset{\sim}{R}^0$ for each $\alpha > 0$, and if $f \in \underset{\sim}{R}^n$ then $U^\alpha f \in \underset{\sim}{R}^{n+1}$, and so (10.1-ii) is verified. It is evident from the explicit construction that $\underset{\sim}{R}$ is the unique minimal cone satisfying (i), (ii), and (iii) of (10.1). Hence (10.1) is proved.

Observe that it is <u>not</u> the case that $\underset{\sim}{C}^+_u \subset \underset{\sim}{R}$. In fact, it might happen that the only continuous elements of $\underset{\sim}{R}$ are the constants.

In the statement of the following lemma we use the notation introduced in the proof of (10.1).

(10.2) LEMMA. (i) $\underset{\sim}{R} = \Lambda[\mathcal{U}(\underset{\sim}{C}_u^+) + \mathcal{U}(\underset{\sim}{R})]$.

(ii) $\Lambda\mathcal{U}(\underset{\sim}{R})$ is uniformly dense in $\underset{\sim}{R}$.

(iii) Each element of $\underset{\sim}{R}$ is β-excessive for some β .

PROOF. Let $\underset{\sim}{H} = \Lambda[\mathcal{U}(\underset{\sim}{C}_u^+) + \mathcal{U}(\underset{\sim}{R})]$. Then as shown in the proof of (10.1), $\underset{\sim}{H}$ is a convex cone. Also by (ii) and (iii) of (10.1), $\underset{\sim}{H} \subset \underset{\sim}{R}$. On the other hand $\underset{\sim}{R}^0 = \mathcal{U}(\underset{\sim}{C}_u^+) \subset \underset{\sim}{H}$. Thus to prove (i) it suffices to show that $\underset{\sim}{R}^{n+1} \subset \underset{\sim}{H}$ whenever $\underset{\sim}{R}^n \subset \underset{\sim}{H}$. But $\mathcal{U}(\underset{\sim}{R}^n) \subset \mathcal{U}(\underset{\sim}{R}) \subset \underset{\sim}{H}$ and so $\underset{\sim}{R}^n + \mathcal{U}(\underset{\sim}{R}^n) \subset \underset{\sim}{H}$ if $\underset{\sim}{R}^n \subset \underset{\sim}{H}$. Therefore $\underset{\sim}{R}^{n+1} = \Lambda(\underset{\sim}{R}^n + \mathcal{U}(\underset{\sim}{R}^n)) \subset \Lambda\underset{\sim}{H} = \underset{\sim}{H}$, proving (i).

If $f = U^\alpha g$ with $g \in \underset{\sim}{C}_u^+$, then $\beta U^{\alpha+\beta} f = \beta U^{\alpha+\beta} U^\alpha g = U^\alpha g - U^{\alpha+\beta} g \to U^\alpha g = f$ uniformly as $\beta \to \infty$. Consequently $\mathcal{U}(\underset{\sim}{R})$ is uniformly dense in $\mathcal{U}(\underset{\sim}{C}_u^+) + \mathcal{U}(\underset{\sim}{R})$, and so using (i), $\Lambda\mathcal{U}(\underset{\sim}{R})$ is uniformly dense in $\underset{\sim}{R}$. Finally (iii) follows from (i) because each element in $\mathcal{U}(\underset{\sim}{C}_u^+) + \mathcal{U}(\underset{\sim}{R})$ is β-excessive for large enough β and because the minimum of two β-excessive functions is again β-excessive. This last fact is an easy consequence of HD2. Indeed, by (2.7) and (2.4) it suffices to show that if f, g are bounded β-excessive functions, then $P_t(f \wedge g) \to (f \wedge g)$ as $t \downarrow 0$. But

$$P_t(f \wedge g)(x) = E^x[f(X_t) \wedge g(X_t)] \to E^x[f(X_0) \wedge g(X_0)] = (f \wedge g)(x)$$

since the minimum of two right continuous functions is right continuous.

We now fix once and for all a countable dense subset $(g_j)_{j \geq 1}$ in $\underset{\sim}{R}$. If $f \in \underset{\sim}{C}_u^+$, $\alpha U^\alpha f \to f$ pointwise as $\alpha \to \infty$ and this implies that $\underset{\sim}{R}$ separates the points of E. Since (g_j) is dense in $\underset{\sim}{R}$, (g_j) also separates the points of E. Let I_j denote the compact interval $[-\|g_j\|, \|g_j\|]$ and let $K = \prod I_j$. Thus K is a compact metrizable space. Of course, the topology of K is the usual product space topology. If $x = (\xi_j)$ and $y = (\eta_j)$ are points in K, then

$$\bar{d}(x, y) = \sum 2^{-j} \frac{|\xi_j - \eta_j|}{1 + |\xi_j - \eta_j|} \tag{10.3}$$

defines a metric on K that is compatible with the product space topology. Next define $\Phi: E \to K$ by $\Phi(x) = (g_j(x))$. Since (g_j) separates the points of E, Φ is injective. Therefore we can carry the metric $\bar{d}$ on K over to E where we call it ρ. That is, for $x, y \in E$

$$\rho(x,y) = \bar{d}[\Phi(x), \Phi(y)] = \sum 2^{-j} \frac{|g_j(x) - g_j(y)|}{1 + |g_j(x) - g_j(y)|} . \tag{10.4}$$

It is obvious that Φ is an isometry of the metric space (E, ρ) onto the metric space $(\Phi(E), \bar{d})$. Therefore Φ extends to an isometry from the completion $\bar{E}$ of E onto $F = \Phi(E)^{-}$, the closure of $\Phi(E)$ in K. We denote the unique extension of ρ from E to $\bar{E}$ by ρ again. Of course, one can identify E with $\Phi(E)$ and $\bar{E}$ with $F = \Phi(E)^{-}$, but perhaps it is clearer to keep them separate for the moment. Since F is compact in K, $(\bar{E}, \rho)$ is a compact metric space and E is a dense subset of $\bar{E}$. Of course, the topology induced by the metric ρ on E is, in general, different from the original topology on E. We shall call this new topology the <u>Ray topology</u> on E. Clearly it depends only on the Ray cone $\underset{\sim}{R}(d)$ and not on the particular sequence (g_j) used in the definition of ρ. However, it appears to depend on the choice of the compact metric space $\hat{E}$ containing E. We shall show in Section 15 that this is not the case; that is, the Ray topology on E is completely determined by the original topology of E and the resolvent (U^{α}). Meanwhile this should cause no difficulty since we shall regard $\hat{E}$ as fixed until Section 15. In the sequel a Ray open set is a set that is open in the Ray topology while an open set is open in the original topology. Similarly a Ray continuous (resp. continuous) function is continuous in the Ray topology (resp. the original topology). With this pattern in mind the reader should have no difficulty with our terminology. Of course, there is only one topology on $\bar{E}$; namely the one induced by ρ. However, for emphasis we shall sometimes refer to the ρ-topology (or ρ-continuous functions) on $\bar{E}$. The compact metric space $\bar{E}$ is called the <u>Ray-Knight compactification</u> of E. However, this is an abuse of language since, in general, $\bar{E}$ depends on the choice of $\hat{E}$.

For each fixed j it follows from (10.4) that

$$2^j \rho(x,y) \geq \frac{|g_j(x) - g_j(y)|}{1 + |g_j(x) - g_j(y)|} \tag{10.5}$$

and from this that if $2^{j+1} \rho(x,y) < 1$, then

$$|g_j(x) - g_j(y)| \leq \frac{2^j \rho(x,y)}{1 - 2^j \rho(x,y)} \leq 2^{j+1} \rho(x,y).$$

Therefore each g_j is ρ-uniformly continuous and hence has a unique extension $\bar{g}_j$ to $\bar{E}$ which is uniformly continuous on $\bar{E}$ and satisfies $\|\bar{g}_j\| = \|g_j\|$. In general, if g is a ρ-uniformly continuous function on E, then we denote its unique uniformly continuous extension to $\bar{E}$ by $\bar{g}$. Note that the map $g \to \bar{g}$ is linear and that $\overline{g \wedge h} = \bar{g} \wedge \bar{h}$. Since (g_j) is dense in $\underset{\sim}{R}$ and each g_j is ρ-uniformly continuous, it is easy to check that each $f \in \underset{\sim}{R}$ is ρ-uniformly continuous, and hence has a unique uniformly continuous extension $\bar{f}$ to $\bar{E}$. Also using a simple approximation argument it is easy to see that if $\bar{x}, \bar{y} \in \bar{E}$, then

$$(10.6) \qquad \rho(\bar{x}, \bar{y}) = \sum 2^{-j} \frac{|\bar{g}_j(\bar{x}) - \bar{g}_j(\bar{y})|}{1 + |\bar{g}_j(\bar{x}) - \bar{g}_j(\bar{y})|} .$$

Let $\bar{\underset{\sim}{R}} = \{\bar{f}: f \in \underset{\sim}{R}\}$, and let $\underset{\sim}{C}(\bar{E})$ denote the space of continuous functions on $\bar{E}$. Then $\bar{\underset{\sim}{R}} \subset \underset{\sim}{C}(\bar{E})$, $\bar{\underset{\sim}{R}}$ is a convex cone, $(\bar{g}_j) \subset \bar{\underset{\sim}{R}}$, and by (10.6) the set $(\bar{g}_j)$ separates the points of $\bar{E}$. Because the map $f \to \bar{f}$ is injective, a function $\bar{f}$ in $\underset{\sim}{C}(\bar{E})$ is in $\bar{\underset{\sim}{R}}$ if and only if its restriction, $\bar{f}|_E$, to E is in $\underset{\sim}{R}$. Since $\bar{f} \wedge \bar{g} = \overline{f \wedge g}$ and $\underset{\sim}{R}$ is closed under "$\wedge$" it follows that $\bar{\underset{\sim}{R}}$ is also. Also $1_{\bar{E}} = \bar{1}_E \in \bar{\underset{\sim}{R}}$. Therefore $\bar{\underset{\sim}{R}} - \bar{\underset{\sim}{R}}$ is a vector space of continuous functions on $\bar{E}$ containing the constants, closed under the pointwise lattice operations, and separating the points of $\bar{E}$. Consequently by the lattice version of the Stone-Weierstrass theorem $\bar{\underset{\sim}{R}} - \bar{\underset{\sim}{R}}$ is uniformly dense in $\underset{\sim}{C}(\bar{E})$.

Next given $f \in \underset{\sim}{R}$ and $\alpha > 0$, $U^\alpha f \in \underset{\sim}{R}$ and so $\overline{U^\alpha f} \in \bar{\underset{\sim}{R}}$. Define $\bar{U}^\alpha$ on $\bar{\underset{\sim}{R}}$ by

$$(10.7) \qquad \bar{U}^\alpha : \bar{f} \to f \to U^\alpha f \to \overline{U^\alpha f}; \; f = \bar{f}|_E .$$

Each step in (10.7) is a cone map (that is, commutes with taking of linear combinations with positive coefficients) and so $\bar{U}^\alpha : \bar{f} \to \overline{U^\alpha f}$ is a cone map from $\bar{\underset{\sim}{R}}$ to $\bar{\underset{\sim}{R}}$. Now define $\bar{U}^\alpha$ on $\bar{\underset{\sim}{R}} - \bar{\underset{\sim}{R}}$ by linearity. It is easy to see that $\bar{U}^\alpha$ is well defined and is a linear map from $\bar{\underset{\sim}{R}} - \bar{\underset{\sim}{R}}$ to $\underset{\sim}{C}(\bar{E})$. If $\bar{f}, \bar{g} \in \bar{\underset{\sim}{R}}$ and $\bar{f} - \bar{g} \geq 0$, then $f - g \geq 0$ and so $U^\alpha f - U^\alpha g \geq 0$. Hence

$$\bar{U}^\alpha(\bar{f} - \bar{g}) = \bar{U}^\alpha \bar{f} - \bar{U}^\alpha \bar{g} = \overline{U^\alpha f} - \overline{U^\alpha g} = (U^\alpha f - U^\alpha g)^- \geq 0 .$$

Therefore $\bar{U}^\alpha$ is a positive, linear map from $\bar{\underset{\sim}{R}} - \bar{\underset{\sim}{R}}$ to $\underset{\sim}{C}(\bar{E})$, and $\|\bar{U}^\alpha\| = \alpha^{-1}$ because $\bar{U}^\alpha 1_{\bar{E}} = \overline{U^\alpha 1_E} = \alpha^{-1} 1_{\bar{E}}$. But $\bar{\underset{\sim}{R}} - \bar{\underset{\sim}{R}}$ is dense in $\underset{\sim}{C}(\bar{E})$ and so each $\bar{U}^\alpha$ extends by continuity to a positive, linear map from $\underset{\sim}{C}(\bar{E})$ to $\underset{\sim}{C}(\bar{E})$ which we again denote by $\overline{U}^{\alpha}$, and $\|\overline{U}^{\alpha}\| = \alpha^{-1}$. Thus, using the Riesz representation theorem, for each $\alpha > 0$ there exists a kernel $\overline{U}^{\alpha}(x, \cdot)$ on $(\overline{E}, \overline{\underline{E}})$ such that $\overline{U}^{\alpha}\overline{f}(x) = \int \overline{U}^{\alpha}(x, dy)\overline{f}(y)$ for all $\overline{f} \in \underset{\sim}{C}(\overline{E})$. Of course, $\overline{\underline{E}}$ is the σ-algebra of Borel subsets of $\overline{E}$. Clearly $\alpha\overline{U}^{\alpha}$ is a Markov kernel for each $\alpha > 0$.

(10.8) PROPOSITION. <u>The family $(\overline{U}^{\alpha})_{\alpha>0}$ is a Ray resolvent on the compact metric space $\overline{E}$.</u>

PROOF. Since each $\overline{U}^{\alpha}$ sends $\underset{\sim}{C}(\overline{E})$ into itself by construction, we need only check that $(\overline{U}^{\alpha})_{\alpha>0}$ satisfies the resolvent equation and that condition (3.1-ii) holds. If $\overline{f} \in \overline{\underset{\sim}{R}}$, then with $f = \overline{f}|_E$

$$\overline{U}^{\alpha}\overline{f} - \overline{U}^{\beta}\overline{f} = (U^{\alpha}f - U^{\beta}f)^{-} = (\beta - \alpha)(U^{\alpha}U^{\beta}f)^{-} .$$

But $U^{\beta}f \in \underset{\sim}{R}$ and so by definition $(U^{\alpha}U^{\beta}f)^{-} = \overline{U}^{\alpha}\overline{(U^{\beta}f)} = \overline{U}^{\alpha}\overline{U}^{\beta}\overline{f}$. Thus the resolvent equation holds on $\overline{\underset{\sim}{R}}$, and this extends to $\overline{\underset{\sim}{R}} - \overline{\underset{\sim}{R}}$ by linearity, and then to $\underset{\sim}{C}(\overline{E})$ by continuity. Therefore $(\overline{U}^{\alpha})_{\alpha>0}$ is a Markov resolvent on $(\overline{E}, \overline{\underline{E}})$. If $f \in \underset{\sim}{R}$, then by (10.2-iii), f is β-excessive for some β. In particular $\alpha U^{\beta+\alpha}f \le f$ for all $\alpha > 0$. But by continuity this gives $\alpha\overline{U}^{\alpha+\beta}\overline{f} \le \overline{f}$. Hence each element of $\overline{\underset{\sim}{R}}$ is continuous and β-supermedian relative to $(\overline{U}^{\alpha})$ for some β. Since $\overline{\underset{\sim}{R}}$ separates the points of $\overline{E}$ it follows that $(\overline{U}^{\alpha})_{\alpha>0}$ is a Ray resolvent on $\overline{E}$.

Let $(\overline{P}_t)$ be the semigroup corresponding to $(\overline{U}^{\alpha})$ constructed in Section 3. In the present situation we have a bit more structure which simplifies certain things.

(10.9) PROPOSITION. <u>The set of degenerate branch points, B_d, of $(\overline{P}_t)$ is empty. The set D of non-branch points of $(\overline{P}_t)$ contains E.</u>

PROOF. Recall from (6.1) that $x \in \overline{E}$ is in B_d if and only if $\overline{P}_0(x, \cdot) = \varepsilon_y$ with $x \ne y$. By (10.2-ii), $\Lambda\mathcal{U}(\underset{\sim}{R})$ is uniformly dense in $\underset{\sim}{R}$, and so $\Lambda\overline{\mathcal{U}}(\overline{\underset{\sim}{R}})$ is dense in $\overline{\underset{\sim}{R}}$. Here $\overline{\mathcal{U}}(\overline{\underset{\sim}{R}})$ stands for all finite sums of the form $\overline{U}^{\alpha_1}\overline{f}_1 + \cdots + \overline{U}^{\alpha_n}\overline{f}_n$ with $\alpha_j > 0$ and $\overline{f}_j \in \overline{\underset{\sim}{R}}$. Since $\overline{\underset{\sim}{R}}$ separates the points of $\overline{E}$, so does $\Lambda\overline{\mathcal{U}}(\overline{\underset{\sim}{R}})$. Observe that if $\overline{U}^{\alpha}\overline{f}(x) = \overline{U}^{\alpha}\overline{f}(y)$ for all $\alpha > 0$ and $\overline{f} \in \overline{\underset{\sim}{R}}$, then $h(x) = h(y)$ for all $h \in \Lambda\overline{\mathcal{U}}(\overline{\underset{\sim}{R}})$. As a result if $\overline{U}^{\alpha}(x, \cdot) = \overline{U}^{\alpha}(y, \cdot)$ for all $\alpha > 0$, then $x = y$. But if $\overline{P}_0(x, \cdot) = \varepsilon_y$, then $\overline{U}^{\alpha}(x, \cdot) = \overline{P}_0\overline{U}^{\alpha}(x, \cdot) = \overline{U}^{\alpha}(y, \cdot)$

for all $\alpha > 0$, and so $x = y$. This establishes the first statement in (10.9).

For the second, let $\overline{f} \in \underset{\sim}{R}$ and $x \in E$. Then because $\overline{f} \in \underset{\sim}{C}(\overline{E})$,

$$\overline{P}_0\overline{f}(x) = \lim_{\alpha \to \infty} \alpha\overline{U}^{\alpha}\overline{f}(x) = \lim_{\alpha \to \infty} \alpha U^{\alpha} f(x) \tag{10.10}$$

where, as usual, f is the restriction of $\overline{f}$ to E. Now f is in $\underset{\sim}{R}$ and hence is β-excessive for some β. Therefore $P_t f(x) \to f(x)$ as $t \to 0$ and so $\alpha U^{\alpha} f(x) \to f(x)$ as $\alpha \to \infty$. Combining this with (10.10) and the fact that $\overline{f}(x) = f(x)$ we see that $\overline{P}_0\overline{f}(x) = \overline{f}(x)$ for each $\overline{f} \in \overline{\underset{\sim}{R}}$. But $\overline{\underset{\sim}{R}} - \overline{\underset{\sim}{R}}$ is dense in $\underset{\sim}{C}(\overline{E})$ and so $\overline{P}_0(x, \cdot) = \varepsilon_x$; that is $x \in D$. Therefore $E \subset D$ proving the second statement in (10.9)

(10.11) REMARK. The second statement in (10.9) is the only place so far that we have used the full power of our hypothesis that X is a right process. The reader may find it of interest to check that everything else is valid if X only satisfies HD1 provided we replace (10.2-iii) by the statement that each element of $\underset{\sim}{R}$ is β-supermedian for some β. Note that this is all that is required in the proof of (10.8).

(10.12) REMARKS AND EXAMPLES

(i) If $U^{\alpha}: \underset{\sim}{C}_u \to \underset{\sim}{C}_u$ for each $\alpha > 0$, then from the explicit construction of $\underset{\sim}{R}$ it is clear that each element of $\underset{\sim}{R}$ is continuous in the original topology. Consequently the Ray topology is coarser than the original topology in this case. Of course, $\alpha U^{\alpha} f \to f$ pointwise as $\alpha \to \infty$ for each $f \in \underset{\sim}{C}_u$. Suppose that not only does $U^{\alpha}: \underset{\sim}{C}_u \to \underset{\sim}{C}_u$ but also $\alpha U^{\alpha} f \to f$ uniformly as $\alpha \to \infty$ for each $f \in \underset{\sim}{C}_u$. In this situation $\underset{\sim}{R} - \underset{\sim}{R}$ is uniformly dense in $\underset{\sim}{C}_u$, and consequently the Ray topology and the original topology coincide. If E itself is a compact metric space in its original topology and $U^{\alpha}: \underset{\sim}{C}(E) \to \underset{\sim}{C}(E)$, then it follows that $\alpha U^{\alpha} f \to f$ uniformly as $\alpha \to \infty$ for each $f \in \underset{\sim}{C}(E)$, and hence the Ray topology and original topology coincide in this case.

(ii) Let X be Brownian motion in $\mathbb{R}^d$ killed when it leaves the unit ball B. Let E be the one point compactification of B and use the point at infinity to make the transition function Markovian as at the end of Section 3. One easily checks that $U^{\alpha}: \underset{\sim}{C}(E) \to \underset{\sim}{C}(E)$, and so by the above discussion the Ray topology and the original topology agree on E and $\overline{E} = E$. Hence the Ray-Knight compactification is quite different from the Martin compactification. See [10].

(iii) Let X be translation to the right at unit speed on the following figure, E:

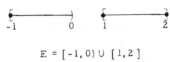

$$E = [-1, 0) \cup [1, 2]$$

with the understanding that when the particle approaches zero it appears at 1 and that 2 is a trap. More explicitly if Y is translation to the right at unit speed on $I = [-1, 1]$ where 1 is a trap and $\varphi: I \to E$ is defined by $\varphi(x) = x$ if $-1 \le x < 0$, $\varphi(x) = x + 1$ if $0 \le x \le 1$, then $X_t = \varphi(Y_t)$. The state space of X is the above figure E and one may take $\hat{E}$ to be $[-1, 2]$. In this case the Ray topology closes the "hole" in E by "identifying 0 and 1." More explicitly the Ray topology is the coarsest topology rendering φ continuous. In this situation the Ray topology on E is strictly coarser than the original topology.

(iv) Let X be translation to the right at unit speed on the following figure, E:

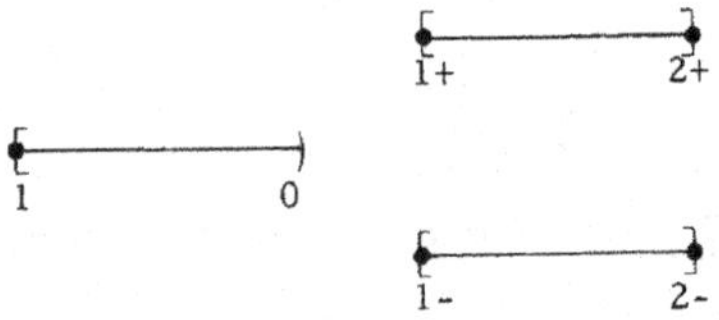

E

where as the particle approaches zero it appears at 1+ with probability 1/2 and at 1- with probability 1/2, and 2+ and 2- are traps. Let $\hat{E}$ be the one point compactification of $\mathbb{R}^2$. In this case $\overline{E}$ is the following figure with the induced Euclidean topology:

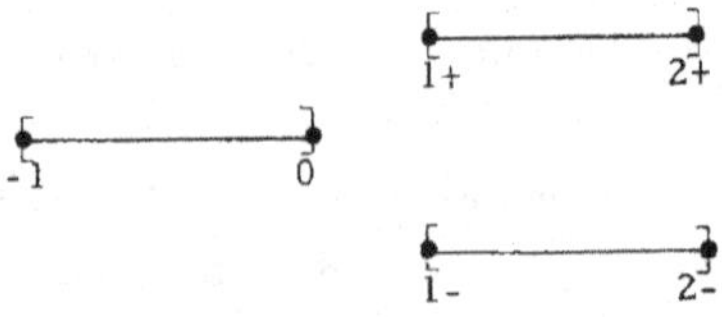

$\overline{E}$

Here 0 is a branch point with $\overline{P}_0(0, \cdot) = \frac{1}{2} \epsilon_{1+} + \frac{1}{2} \epsilon_{1-}$. Note that the original and Ray topologies agree on E in this case.

(v) Let X be translation to the right on $I = [-1, 1]$ at unit speed killed with probability 1/2 as it passes through the origin and with 1 a trap. Let $E = I \cup \{\Delta\}$ where $\Delta = (0, 1) \in \mathbb{R}^2$ is used to render the transition function

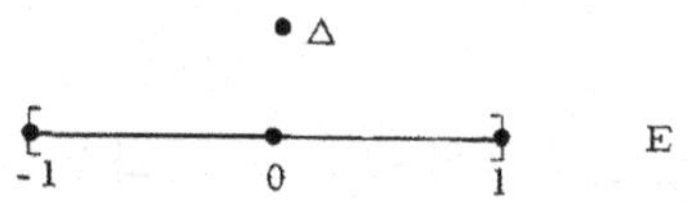

Markovian. In this case $\overline{E}$ looks as follows:

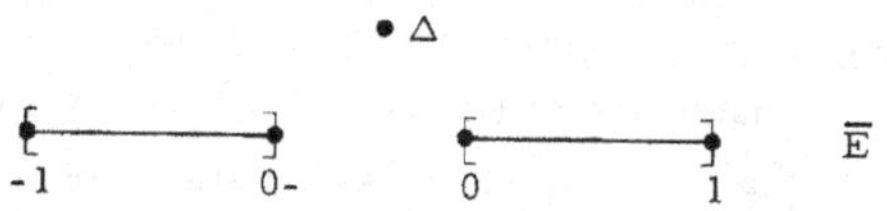

with the induced Euclidean topology. Here the point 0 has been "ramified" into two points 0- and 0. The point 0- is a branch point with $\overline{P}_0(0-, \cdot) = 1/2\ \epsilon_\Delta + 1/2\ \epsilon_0$. In this case the Ray topology on E is strictly finer than the original topology. Of course, this does <u>not</u> contradict (i) since U^α does not send $\underset{\sim}{C}_u$ into $\underset{\sim}{C}_u$ in this case.

(vi) By combining (iii) and (v) we see that, in general, the Ray topology and the original topology are not comparable.

11. COMPARISON OF PROCESSES

This section is a continuation of Section 10. The assumptions and notations are exactly those of Section 10. In particular X is our fixed right process with state space E, $\overline{E}$ is the completion of E in the ρ metric, $(\overline{U}^\alpha)$ is the Ray resolvent on $\overline{E}$ constructed in (10.8), and $(\overline{P}_t)$ is the corresponding semigroup. Let Y be the Ray process on $\overline{E}$ associated with $(\overline{P}_t)$ and $(\overline{U}^\alpha)$ as in Section 5.

Let us review the notation for the two processes X and Y. For X, Ω is the set of all right continuous functions ω from $\mathbb{R}^+$ to E where we use the original topology on E. The σ-algebras $\underline{\underline{F}}^0$, $\underline{\underline{F}}^{0*}$, $\underline{\underline{F}}^\mu$, etc. are defined as in Section 9, and for each probability μ on $(E, \underline{\underline{E}})$, P^μ denotes the corresponding probability on $(\Omega, \underline{\underline{F}}^\mu)$ under which X is a Markov process with initial measure μ and transition function P_t. For the Ray process Y, W denotes the set of all right continuous functions w from $\mathbb{R}^+$ to D that have left limits in $\overline{E}$ where, of course, D and $\overline{E}$ are equipped with the ρ-topology. Here D is the set of non-branch points of $(\overline{P}_t)$ and by (10.9) we know that $E \subset D \subset \overline{E}$. (Let us agree that if f is a function on $\mathbb{R}^+$ to some topological space, then the statement that f has left limits means that it has a left limit at each point of $\mathbb{R}^{++} = (0, \infty)$, and by definition we put $f(0-) = f(0)$.) The σ-algebras $\underline{\underline{G}}^0$, $\underline{\underline{G}}^\mu$, etc. are defined as in Section 5. If μ is a probability on $(\overline{E}, \overline{\underline{\underline{E}}})$, then $\overline{P}^\mu$ denotes the corresponding probability on $(W, \underline{\underline{G}}^\mu)$ under which Y is a Markov process with initial measure $\mu\overline{P}_0$ and transition function $\overline{P}_t$. Note that relative to the process Y, $\overline{E}$ now plays the role of E in Sections 5, 6 and 7.

Let us emphasize that the object of interest to us is the right process X and that the Ray process Y is regarded as a tool to aid in the study of X. The main goal of this section is to make precise and prove the following statement:

(11.1) <u>For each μ on E, the measures P^μ and $\overline{P}^\mu$ are the same.</u>

Once this is established we will be able to apply all of the results of Sections 5,

6, and 7 proved for Y to the right process X. However, there are several difficulties to overcome before (11.1) makes any sense. First of all P^μ, resp. $\overline{P}^\mu$, is defined for μ a probability on $(E, \underline{\underline{E}})$, resp. on $(\overline{E}, \overline{\underline{\underline{E}}})$. Therefore we first need some information about the relationship between these measurable spaces. Of course, E is a subset of $\overline{E}$, but we need to know how the set E sits in $\overline{E}$. It is not true, in general, that E is Borel in $\overline{E}$; that is, $E \in \overline{\underline{\underline{E}}}$, but it turns out that E is <u>universally measurable</u> in $\overline{E}$, and that the trace of $\overline{\underline{\underline{E}}}^*$ on E is $\underline{\underline{E}}^*$. This is proved in (11.3).

Having accomplished this, given a probability μ on $(E, \underline{\underline{E}}^*)$ we may regard μ as a measure on $(\overline{E}, \overline{\underline{\underline{E}}}^*)$ that is carried by E and so we can at least define P^μ and $\overline{P}^\mu$. But P^μ is defined on $(\Omega, \underline{\underline{F}}^\mu)$ while $\overline{P}^\mu$ is defined on $(W, \underline{\underline{G}}^\mu)$, and so it still does not make sense to talk about the equality of these measures. Let Ω_0 be the set of those $\omega \in \Omega$ such that $t \to \omega(t)$ is right Ray continuous and has left ρ-limits in $\overline{E}$. Since $E \subset D$, it is clear that $\Omega_0 \subset W$ and that $X_t = Y_t$ on Ω_0. It is immediate from the very definition of the ρ-topology that $\Omega_0 \in \underline{\underline{F}}$ and $P^\mu(\Omega_0) = 1$ for each μ. See (11.2). We then show that for each μ on $(E, \underline{\underline{E}}^*)$, Ω_0 has $\overline{P}^\mu$ outer measure one. Thus the trace of $\overline{P}^\mu$ on Ω_0 is a probability measure, and it is <u>this measure</u> that agrees with P^μ restricted to Ω_0. It is in this sense that (11.1) has a meaning. This result is contained in (11.8). There is one more crucial fact. Namely that in discussing the left limits of X in the ρ-topology the only points of $\overline{E} - E$ that play any role are the branch points of the semigroup $(\overline{P}_t)$. This is (11.14).

After all this talk let us get down to work. Recall that (g_j) is our fixed countable dense subset of the Ray cone $\underset{\sim}{R}$ that is used in defining the metric ρ. Since $E \subset \overline{E}$ we may regard $t \to X_t(\omega)$ as a map from $\mathbb{R}^+$ to $\overline{E}$ for each ω. Let μ be a probability on $(E, \underline{\underline{E}})$. Each g_j is β-excessive relative to (P_t) for some $\beta > 0$ by (10.2-iii), and so by HD2 and (9.6) almost surely P^μ, $t \to g_j(X_t)$ is right continuous and has left limits. Thus from the very definition of ρ in (10.4) it follows that almost surely P^μ, $t \to X_t$ is right Ray continuous and has left ρ-limits in $\overline{E}$. Since μ is arbitrary we have proved the following result.

(11.2) PROPOSITION. <u>Let</u> Ω_0 <u>denote the set of those</u> ω <u>in</u> Ω <u>such that</u> $t \to \omega(t)$ <u>is right Ray continuous and has left</u> ρ<u>-limits in</u> $\overline{E}$. <u>Then</u> $\Omega_0 \in \underline{\underline{F}}$ <u>and</u> $P^\mu(\Omega_0) = 1$ <u>for each</u> μ <u>on</u> $(E, \underline{\underline{E}})$.

We now come to the fact that E is universally measurable in $\bar{E}$. This is the key technical fact of this section and we formulate it as a proposition. For its statement recall the definition of K, F, and the map $\Phi: E \to F \subset K$ given after the proof of (10.2), and also from (9.7) that (P_t) is Borel if E is Lusinien and $P_t: b\underline{\underline{E}} \to b\underline{\underline{E}}$.

(11.3) PROPOSITION. (i) *If (P_t) is Borel, then E is Borel in $\bar{E}$, that is $E \in \bar{\underline{\underline{E}}}$. In this case $\underline{\underline{E}}$ is the trace of $\bar{\underline{\underline{E}}}$ on E, and $\Phi: E \to F$ is $\underline{\underline{E}} | \underline{\underline{B}}(F)$ measurable where $\underline{\underline{B}}(F)$ is the σ-algebra of Borel subsets of F.*

(ii) In the general case E is universally measurable in $\bar{E}$, that is, $E \in \bar{\underline{\underline{E}}}^*$, and $\underline{\underline{E}}^*$ is the trace of $\bar{\underline{\underline{E}}}^*$ on E. In this case Φ is $\sigma(\underset{\sim}{R}) | \underline{\underline{B}}(F)$ measurable where $\sigma(\underset{\sim}{R})$ is the σ-algebra on E generated by $\underset{\sim}{R}$. Furthermore if $\underline{\underline{E}}_r$ is the trace of $\bar{\underline{\underline{E}}}$ on E, then $\underline{\underline{E}} \subset \underline{\underline{E}}_r = \sigma(\underset{\sim}{R}) \subset \underline{\underline{E}}^*$ and $\underline{\underline{E}}_r^* = \underline{\underline{E}}^*$.

PROOF. Let (g_j) be our fixed countable dense subset of $\underset{\sim}{R}$. Then $K = \prod I_j$ where $I_j = [-\|g_j\|, \|g_j\|]$ and $\Phi(x) = (g_j(x))$. A base for the (product) topology of K consists of sets of the form

$$G = G_1 \times \cdots \times G_n \times I_{n+1} \times \cdots$$

where each G_j is open in I_j. Thus $\Phi^{-1}(G) = \bigcap_{j=1}^{n} g_j^{-1}(G_j)$, and so Φ is $\sigma(\underset{\sim}{R}) | \underline{\underline{K}}$ measurable where $\underline{\underline{K}}$ is the σ-algebra of Borel subsets of K. Since F is closed in K, Φ is $\sigma(\underset{\sim}{R}) | \underline{\underline{B}}(F)$ measurable. If (P_t) is Borel, then $U^\alpha: b\underline{\underline{E}} \to b\underline{\underline{E}}$ for each $\alpha > 0$. Consequently by the explicit construction of $\underset{\sim}{R}$ each f in $\underset{\sim}{R}$ is $\underline{\underline{E}}$ measurable, and so $\sigma(\underset{\sim}{R}) \subset \underline{\underline{E}}$ when (P_t) is Borel. In the general case if $f \in \underset{\sim}{C}_u^+$, then $U^\alpha f \in \underset{\sim}{R}$ and $\alpha U^\alpha f \to f$ pointwise as $\alpha \to \infty$. This implies that $\underset{\sim}{C}_u \subset b\sigma(\underset{\sim}{R})$, and hence by (8.6), $\underline{\underline{E}} \subset \sigma(\underset{\sim}{R})$. So far we have proved the two assertions about Φ and that $\underline{\underline{E}} \subset \sigma(\underset{\sim}{R})$ with equality if (P_t) is Borel.

In Section 10, Φ was used in two senses: as the injection of E into K, and also as the isometry from $\bar{E}$ onto F, the closure of $\Phi(E)$ in K. For clarity let us denote this isometry by Φ^* during the present proof. Of course, the restriction of Φ^* to E is Φ, and so $i = (\Phi^*)^{-1} \circ \Phi$ is the inclusion map of E in $\bar{E}$.

Now suppose that (P_t) is Borel. In this case E is a Borel subset of the compact metric space $\hat{E}$ and we have shown that Φ is a Borel measurable injection of E into F. Therefore if $B \in \underline{\underline{E}}$, then B is Borel in $\hat{E}$ and Φ restricted to B is injective and Borel measurable. Hence by (8.7), $\Phi(B)$ is Borel in F. Since Φ^* is an isometry from $\overline{E}$ onto F both Φ^* and $(\Phi^*)^{-1}$ are Borel measurable mappings. But $B = i(B) = (\Phi^*)^{-1}[\Phi(B)]$, and so it follows that $B \in \overline{\underline{\underline{E}}}$; that is, B is Borel in $\overline{E}$. In particular, $E \in \overline{\underline{\underline{E}}}$ and so the trace, $\underline{\underline{E}}_r$, of $\overline{\underline{\underline{E}}}$ on E consists of those subsets $A \subset E$ such that $A \in \overline{\underline{\underline{E}}}$. It is immediate from what was proved above that $\underline{\underline{E}} \subset \underline{\underline{E}}_r$. Conversely suppose that $B \subset E$ and $B \in \overline{\underline{\underline{E}}}$. Then $B = i^{-1}(B) = \Phi^{-1}[\Phi^*(B)]$, and because $\Phi^*(B) \in \underline{\underline{B}}(F)$ this gives $B \in \underline{\underline{E}}$. Therefore $\underline{\underline{E}}_r = \underline{\underline{E}}$. This completes the proof of (11.3-i).

We next turn to the last assertion in (11.3-ii). Since we already know that $\underline{\underline{E}} \subset \sigma(\underset{\sim}{R}) \subset \underline{\underline{E}}^*$, it suffices to show that $\underline{\underline{E}}_r = \sigma(\underset{\sim}{R})$. Let $\underset{\sim}{C}_u(E, \rho) = \underset{\sim}{C}_\rho$ be the restrictions of the functions in $\underset{\sim}{C}(\overline{E})$ to E. Then $\underset{\sim}{R} - \underset{\sim}{R}$ is a dense subset of $\underset{\sim}{C}_\rho$ because $\overline{\underset{\sim}{R}} - \overline{\underset{\sim}{R}}$ is dense in $\underset{\sim}{C}(\overline{E})$. Consequently $\sigma(\underset{\sim}{R}) = \sigma(\underset{\sim}{C}_\rho)$. But by (8.4) applied to $(\overline{E}, \rho)$, one has $\sigma(\underset{\sim}{C}_\rho) = \underline{\underline{E}}_r$ and so $\sigma(\underset{\sim}{R}) = \underline{\underline{E}}_r$. This establishes the last assertion in (11.3-ii) since $\underline{\underline{E}} \subset \underline{\underline{E}}_r \subset \underline{\underline{E}}^*$ implies that $\underline{\underline{E}}_r^* = \underline{\underline{E}}^*$.

The next step in the proof of (11.3-ii) is to show that if μ is a finite measure on $(E, \underline{\underline{E}})$, then there exists a set $E_\mu \subset E$ with E_μ in <u>both</u> $\underline{\underline{E}}$ and $\overline{\underline{\underline{E}}}$ such that $\mu(E) = \mu(E_\mu)$. To this end fix such a μ. We extend μ to $(E, \underline{\underline{E}}^*)$ as usual, and then according to (8.4-iii) we may regard μ as a measure on $(\hat{E}, \hat{\underline{\underline{E}}}^*)$ that is carried by E. Since E is universally measurable in $\hat{E}$ there exists $E_\mu^0 \subset E$ with $E_\mu^0 \in \hat{\underline{\underline{E}}}$ and $\mu(E_\mu^0) = \mu(E)$. By (8.4-i), E_μ^0 is also in $\underline{\underline{E}}$. Each g_j is $\underline{\underline{E}}^*$ measurable and so there exist $g_j', g_j'' \in b\underline{\underline{E}}^+$ with $g_j' \le g_j \le g_j''$ and $\mu(g_j'' - g_j') = 0$. Replacing g_j'' by $g_j'' \wedge \|g_j\|$ we may assume that $\|g_j'\| \le \|g_j''\| \le \|g_j\|$. Again by (8.4-i) for each j there exist $h_j', h_j'' \in b\hat{\underline{\underline{E}}}^+$ such that $g_j' = h_j'|_E$ and $g_j'' = h_j''|_E$. Clearly we may assume that $\|h_j'\| \le \|h_j''\| \le \|g_j\|$. Since μ is carried by E, $\mu(\{h_j' \ne h_j''\}) = \mu(g_j'' - g_j') = 0$. Define $E_\mu^1 = \cap_j \{h_j' = h_j''\}$ and $E_\mu = E_\mu^1 \cap E_\mu^0$. Then $E_\mu \subset E$, $E_\mu \in \hat{\underline{\underline{E}}}$, and $\mu(E_\mu) = \mu(E)$. Let $\Psi\colon E_\mu \to K$ be defined by $\Psi(x) = (h_j'(x))$. Since $\|h_j'\| \le \|g_j\|$, it is clear that Ψ does indeed map E_μ into K. Also by the definition of E_μ, Ψ is equal to $\Phi|_{E_\mu}$. Consequently Ψ is injective, and since each $h_j' \in b\hat{\underline{\underline{E}}}$ and $E_\mu \in \hat{\underline{\underline{E}}}$ it follows that Ψ is a (Borel) measurable mapping from E_μ to K, that is if L is a Borel subset of K, then $\Psi^{-1}(L) \in \hat{\underline{\underline{E}}}$ and, of course, $\Psi^{-1}(L) \subset E_\mu$. Therefore by (8.7), $\Phi(E_\mu) = \Psi(E_\mu)$ is Borel in K, and hence in F. As before $E_\mu = i(E_\mu) = (\Phi^*)^{-1}[\Phi(E_\mu)]$, and so E_μ is Borel in $\overline{E}$, that is, $E_\mu \in \overline{\underline{\underline{E}}}$. Also $E_\mu \in \underline{\underline{E}}$ because of (8.4-i). We have now established the assertion at the beginning of this paragraph.

Next let λ be a finite measure on $(\overline{E}, \overline{\underline{E}})$ and let μ be the trace of λ on $(E, \underline{\underline{E}}_r)$ constructed in (8.3). Let $E_\mu \subset E$ be defined as above relative to this μ (restricted to $\underline{\underline{E}}$ to be precise). Then $E_\mu \in \underline{\underline{E}} \subset \underline{\underline{E}}_r$, $\mu(E_\mu) = \mu(E)$, and most importantly $E_\mu \in \overline{\underline{\underline{E}}}$. As in (8.3) choose $A_0 \supset E$, $A_0 \in \overline{\underline{\underline{E}}}$ with minimal λ measure so that $\mu(B) = \lambda(A \cap A_0)$ whenever $B = A \cap E$ with $A \in \overline{\underline{\underline{E}}}$. Now $E_\mu \in \underline{\underline{E}} \subset \underline{\underline{E}}_r$ and so $\mu(E_\mu) = \lambda(E_\mu \cap A_0) = \lambda(E_\mu)$. Moreover, $\lambda(A_0) = \mu(E) = \mu(E_\mu)$. Therefore $E_\mu \subset E \subset A_0$, E_μ and A_0 are in $\overline{\underline{\underline{E}}}$, and $\lambda(E_\mu) = \lambda(A_0)$. Since λ is arbitrary this says that $E \in \overline{\underline{\underline{E}}}^*$. It now follows readily from (8.5) and $\underline{\underline{E}} \subset \underline{\underline{E}}_r \subset \underline{\underline{E}}^*$ that the trace of $\overline{\underline{\underline{E}}}^*$ on E is $\underline{\underline{E}}_r^* = \underline{\underline{E}}^*$, completing the proof of (11.3).

REMARK. In view of (8.4-i), the σ-algebra $\underline{\underline{E}}_r$ is the σ-algebra of Borel subsets of the metric space (E, ρ), that is, the Ray Borel subsets of E. Of course, we now know that E is a U-space in the Ray topology. At the end of Section 10 we pointed out that the original and Ray topologies on E are not comparable. However, the Ray Borel structure $\underline{\underline{E}}_r$ is <u>always</u> finer than the original Borel structure $\underline{\underline{E}}$ since $\underline{\underline{E}} \subset \underline{\underline{E}}_r$. Moreover both topologies give rise to the <u>same</u> universal structure; that is, $\underline{\underline{E}}^* = \underline{\underline{E}}_r^*$.

We now are ready to compare the processes X and Y. Given a probability μ on $(E, \underline{\underline{E}})$, we extend it to $(E, \underline{\underline{E}}^*)$ as usual. By (11.3) we may now regard μ as a measure on $(\overline{E}, \overline{\underline{\underline{E}}}^*)$ that is carried by E, and so we can construct the measure $\overline{P}^\mu$ on $(W, \underline{\underline{G}}^0)$ associated with the Ray process Y. We also have the measure P^μ on $(\Omega, \underline{\underline{F}})$ associated with X.

If $\overline{f} \in \overline{\underset{\sim}{R}}$, then $f = \overline{f}|_E \in \underset{\sim}{R}$ and by definition $\overline{U}^\alpha \overline{f} = U^\alpha f$ on E. Since $X_t(\omega) \in E$ for all t and ω, one has for each $\alpha > 0$ and $t \geq 0$

$$e^{-\alpha t} \overline{U}^\alpha \overline{f}(X_t) = e^{-\alpha t} U^\alpha f(X_t) = e^{-\alpha t} E^\mu \left[\int_0^\infty e^{-\alpha s} f(X_{t+s}) ds \mid \underline{\underline{F}}_t^\mu \right]$$

$$= e^{-\alpha t} E^\mu \left[\int_0^\infty e^{-\alpha s} \overline{f}(X_{t+s}) ds \mid \underline{\underline{F}}_t^\mu \right] .$$

For $\Lambda \in \underline{\underline{F}}_t^\mu$, using the relationship between $(\overline{U}^\alpha)$ and $(\overline{P}_t)$ this yields

$$(11.4) \qquad \int_0^\infty e^{-\alpha s} E^\mu[\bar{P}_s \bar{f}(X_t); \Lambda] ds = \int_0^\infty e^{-\alpha s} E^\mu[\bar{f}(X_{t+s}); \Lambda] ds .$$

But $s \to \bar{P}_s \bar{f}$ is right continuous since $\bar{f} \in \underset{\sim}{C}(\bar{E})$, while by (11.2) the map $t \to X_t$ from $\mathbb{R}^+$ to $(\bar{E}, \rho)$ is P^μ almost surely right continuous and so $s \to E^\mu[\bar{f}(X_{t+s}); \Lambda]$ is right continuous. Therefore (11.4) and the uniqueness theorem for Laplace transforms imply that

$$(11.5) \qquad E^\mu[\bar{f}(X_{t+s}) | \underline{\underline{F}}^\mu_t] = \bar{P}_s \bar{f}(X_t)$$

for $t, s \geq 0$ and $\bar{f} \in \bar{R}$. Since $\underset{\sim}{\bar{R}} - \underset{\sim}{\bar{R}}$ is dense in $\underset{\sim}{C}(\bar{E})$ it follows that (11.5) holds for $\bar{f} \in b\bar{\underline{\underline{E}}}$. Note that if $\bar{g} \in b\bar{\underline{\underline{E}}}$, then by (11.3), $g = \bar{g}|_E$ is in $b\underline{\underline{E}}^*$, and so $\bar{g}(X_t) = g(X_t)$ is $\underline{\underline{F}}_t$ measurable. In particular the right side of (11.5) is $\underline{\underline{F}}_t$ measurable. Now we may regard (X_t) as a process with values in $(\bar{E}, \bar{\underline{\underline{E}}})$ over $(\Omega, \underline{\underline{F}}^\mu, \underline{\underline{F}}^\mu_t, P^\mu)$, and as such in view of (11.2) and (11.5) it has the following properties:

(11.6)
- (i) Almost surely P^μ it is right Ray continuous and has left limits in $\bar{E}$ (in the ρ-topology).
- (ii) It is Markov with transition semigroup $(\bar{P}_t)$.
- (iii) It is carried by E and has initial measure $\mu = \mu\bar{P}_0$ since μ is carried by E.

Let Ω_0 be the set of all maps from $\mathbb{R}^+$ to E that are right continuous for the original topology on E and which are also right Ray continuous and have left limits in $\bar{E}$ for the ρ-topology on $\bar{E}$. Since $E \subset D$ it is clear that $\Omega_0 \subset \Omega \cap W$ and that for each $t, X_t = Y_t$ on Ω_0. By (11.2), $\Omega_0 \in \underline{\underline{F}}$ and $P^\mu(\Omega_0) = 1$ for each probability μ on $(E, \underline{\underline{E}})$. Fix such a μ. We now regard each X_t as a map from Ω_0 to $(\bar{E}, \bar{\underline{\underline{E}}}^*)$ and let $\underline{\underline{F}}^*$ be the σ-algebra on Ω_0 generated by these maps. Let $(\mu_J: J \in \Phi(\mathbb{R}^+))$ be the projective system over $(\bar{E}, \bar{\underline{\underline{E}}}^*)$ of the finite dimensional distributions of the process $X^\mu = (\Omega_0, \underline{\underline{F}}^*, X_t, P^\mu)$. (Here $\Phi(\mathbb{R}^+)$ denotes the collection of all nonempty finite subsets of $\mathbb{R}^+$ as in Section 8.) Clearly $X_J(\Omega_0) = E^J$ and since if $J = (t_1, \ldots, t_n)$,

$$\mu_J(\bar{E}^J - E^J) = P^\mu[(X_{t_1}, \ldots, X_{t_n}) \in \bar{E}^J - E^J] = 0 ,$$

it follows that each X_J as a map from Ω_0 to $(\bar{E}^J, \bar{\underline{\underline{E}}}^{*J}, \mu_J)$ is almost surjective. If

$$(11.7) \qquad \underline{\underline{A}} = \{\Lambda \subset \Omega_0 : \Lambda = X_j^{-1}(B_J)\,;\ B_J \in \overline{\underline{\underline{E}}}^{*J},\ J \in \Phi(\mathbb{R}^+)\}\ ,$$

then the finitely additive measure Q on $\underline{\underline{A}}$ constructed in (8.9) is just the restriction of P^{μ} to $\underline{\underline{A}}$ and is countably additive. (This also is a consequence of (8.11).)

Next consider the process $Y^{\mu} = (W, \underline{\underline{G}}^{0*}, Y_t, \overline{P}^{\mu})$ where each Y_t is regarded as a map from W to $(\overline{E}, \overline{\underline{\underline{E}}}^*)$ and $\underline{\underline{G}}^{0*}$ is the σ-algebra generated by these maps. Let $(\overline{\mu}_J : J \in \Phi(\mathbb{R}^+))$ be the projective system over $(\overline{E}, \overline{\underline{\underline{E}}}^*)$ of the finite dimensional distributions of Y^{μ}. Since μ is carried by $E \subset D$ the initial measure of Y^{μ} is $\mu\overline{P}_0 = \mu$. Therefore according to (11.6) both X^{μ} and Y^{μ} may be viewed as Markov processes with state space $(\overline{E}, \overline{\underline{\underline{E}}}^*)$, transition semigroup $(\overline{P}_t)$, and initial measure μ. Consequently they have the same finite dimensional distributions; that is, $\mu_J = \overline{\mu}_J$ for each $J \in \Phi(\mathbb{R}^+)$. Finally let Y_J^0 be the restriction of Y_J to Ω_0. Then $Y_J^0 = X_J$ and consequently each Y_J^0 as a map from Ω_0 to $\overline{E}^J$ is almost surjective. Since $\underline{\underline{A}}$ defined in (11.7) is unchanged if X_J is replaced by Y_J^0, the finitely additive measure $\overline{Q}$ on $\underline{\underline{A}}$ constructed from $(\overline{\mu}_J : J \in \Phi(\mathbb{R}^+))$ is equal to Q and hence is countably additive. Therefore by (8.11), Ω_0 has $\overline{P}^{\mu}$ outer measure one. Clearly the trace of $\underline{\underline{G}}^{0*}$ on Ω_0 is $\underline{\underline{F}}^* = \sigma(\underline{\underline{A}})$, and it follows from (8.11) that the trace of $\overline{P}^{\mu}$ on $(\Omega_0, \underline{\underline{F}}^*)$ is P^{μ}.

We now change our notation and let Ω be all maps from $\mathbb{R}^+$ to E that are right continuous in both the original and the Ray topology and which have left limits in $\overline{E}$ in the ρ-topology. (This is what we called Ω_0 in the preceding paragraphs.) Let $X_t(\omega) = \omega(t)$ and note that Ω is closed under the shifts θ_t defined by $\theta_t\omega(s) = \omega(t+s)$. Define $\underline{\underline{F}}^{0*} = \sigma(X_s ; s \geq 0)$ and $\underline{\underline{F}}_t^{0*} = \sigma(X_s ; s \leq t)$ where each X_s is regarded as a map from Ω to $(\overline{E}, \overline{\underline{\underline{E}}}^*)$, or equivalently to $(E, \underline{\underline{E}}^*)$. We summarize what has been proved in the preceding paragraphs in the following:

(11.8) THEOREM. *For each probability μ on $(E, \underline{\underline{E}})$ let P^{μ} be the measure on $(\Omega, \underline{\underline{F}}^{0*})$ constructed from μ and (P_t) and let $\overline{P}^{\mu}$ be the measure on $(\Omega, \underline{\underline{F}}^{0*})$ constructed from μ and $(\overline{P}_t)$ — these constructions are possible by the above considerations. Then $P^{\mu} = \overline{P}^{\mu}$ and $(X_t, \underline{\underline{F}}_t^{0*}, P^{\mu})$ is a Markov process with initial measure μ and having (P_t), resp. $(\overline{P}_t)$, as transition semigroup if one takes $(E, \underline{\underline{E}}^*)$, resp. $(\overline{E}, \overline{\underline{\underline{E}}}^*)$, as state space. More precisely if $f \in b\underline{\underline{E}}^*$ and*

$\bar{f} \in b\bar{\underline{\underline{E}}}^*$, then

$$E^\mu[f(X_{t+s})|\underline{\underline{F}}_s^{0*}] = P_t f(X_s)$$

$$E^\mu[\bar{f}(X_{t+s})|\underline{\underline{F}}_s^{0*}] = \bar{P}_t\bar{f}(X_s)$$

for all $t, s \in \mathbb{R}^+$.

We identify $b\underline{\underline{E}}^*$ with those elements of $b\bar{\underline{\underline{E}}}^*$ which vanish on $\bar{E} - E$. An immediate consequence of (11.8) is that for all $f \in b\underline{\underline{E}}^*$

$$P_t f(X_s) = \bar{P}_t f(X_s)$$

almost surely P^μ. Taking $\mu = \epsilon_x$ with $x \in E$ and $s = 0$ we obtain the following:

(11.9) COROLLARY. *If* $x \in E$, $P_t(x, \cdot) = \bar{P}_t(x, \cdot)$ *for all* $t \geq 0$ *and* $U^\alpha(x, \cdot) = \bar{U}^\alpha(x, \cdot)$ *for all* $\alpha > 0$. *In particular* $\bar{P}_t(x, \cdot)$ *and* $\bar{U}^\alpha(x, \cdot)$ *are carried by* E *if* $x \in E$.

The results of (11.8) and (11.9) are the justification of the statement (11.1). They give us a way of applying what we know about Ray processes, that is the measures $\bar{P}^\mu$, to right processes, that is the measures P^μ. Of course, as long as μ is carried by E there is no longer any need to distinguish these measures. The σ-algebras $\underline{\underline{F}}^\mu$, $\underline{\underline{F}}$, $\underline{\underline{F}}_t^\mu$, and $\underline{\underline{F}}_t$ are defined as in Section 9 ($\underline{\underline{F}}$ and $\underline{\underline{F}}_t$ are the intersections over all μ on E of $\underline{\underline{F}}^\mu$ and $\underline{\underline{F}}_t^\mu$ respectively), and all of the results of Section 9 are valid on the current Ω. Also, as long as μ is carried by E, the results of Sections 5, 6, and 7 are valid for the process $(\Omega, \underline{\underline{F}}^\mu, \underline{\underline{F}}_t^\mu, X_t, P^\mu)$. We adopt the convention that two functions on $\mathbb{R}^+ \times \Omega$ are called *indistinguishable* provided they are P^μ indistinguishable for all μ on E.

(11.10) NOTATION. *For* $t > 0$ *and* $\omega \in \Omega$ *let* $X_{t-}(\omega)$ *denote the left limit in the* ρ*-topology of* $s \to X_s(\omega)$ *at* t. *Thus* $X_{t-}(\omega)$ *always exists as a point of* $\bar{E}$. *Let* $X_{0-}(\omega) = X_0(\omega)$. *Let* $X_{t-}^*(\omega)$ *denote the left limit in the original topology of* E *of* $s \to X_s(\omega)$ *at* t *whenever it exists. Let* $X_{0-}^*(\omega) = X_0(\omega)$. *Of course, whenever* $X_{t-}^*(\omega)$ *exists, it is a point of* E.

The next proposition is the crucial fact needed for studying the process $X^- = (X_{t-})$. In its statement we write $f \circ X^-$ for the process $(f \circ X_{t-})$ for any function f on $\bar{E}$. Also remember that $\underline{\underline{E}} \subset \underline{\underline{E}}_r$.

(11.11) PROPOSITION. (i) Let $f \in b\bar{\underline{\underline{E}}}^*$ with $f = 0$ on E. Then for each μ on E, the process $f \circ X^-$ is P^μ indistinguishable from a previsible process over $(\Omega, \underline{\underline{F}}^\mu, \underline{\underline{F}}_t^\mu, P^\mu)$.

(ii) If $f \in b\underline{\underline{E}}_r$, then for each μ on E, $f \circ X^-$ is P^μ indistinguishable from a previsible process over $(\Omega, \underline{\underline{F}}^\mu, \underline{\underline{F}}_t^\mu, P^\mu)$.

REMARK. In (ii), f is extended to $\bar{E}$ by setting $f = 0$ on $\bar{E} - E$.

PROOF. Fix μ on E and all statements will be relative to the system $(\Omega, \underline{\underline{F}}^\mu, \underline{\underline{F}}_t^\mu, P^\mu)$. Let

$$J = \{X \neq X^-\} = \{(t, \omega) \colon X_t(\omega) \neq X_{t-}(\omega)\} .$$

By D-IV-T30 there exists a sequence (T_n) of stopping times with disjoint graphs such that $J = \cup [[T_n]]$. Note that each $T_n > 0$ since $X_0 = X_{0-}$ by definition. For each n let $\nu_n(dx) = P^\mu(X_{T_n-} \in dx,\ T_n < \infty)$ be the distribution of X_{T_n-} under P^μ, and let $\nu = \sum 2^{-n} \nu_n$. Then ν is a finite measure on $(\bar{E}, \bar{\underline{\underline{E}}}^*)$.

Let $f \in b\bar{\underline{\underline{E}}}^*$ with $f = 0$ on E. Since $f^+ = f \vee 0$ and $f^- = -(f \wedge 0)$ both vanish on E if f does, we may, and shall, assume that $f \geq 0$ in proving (i). Since $f \in b\bar{\underline{\underline{E}}}_+^*$, there exists $g \in b\bar{\underline{\underline{E}}}_+$ with $0 \leq g \leq f$ and $\nu(f - g) = 0$ where ν is the measure defined just above. Note that $g = 0$ on E also. Now

$$f \circ X^- = g \circ X^- + (f \circ X^- - g \circ X^-)$$

and $g \circ X^-$ is previsible since $g \in b\bar{\underline{\underline{E}}}$. Thus to prove (i) it suffices to show that $\{g \circ X^- \neq f \circ X^-\}$ is evanescent. Since both f and g vanish on E and $X_t(\omega) \in E$ for all t, ω, we have

$$\{g \circ X^- \neq f \circ X^-\} = \{g \circ X^- < f \circ X^-\} \subset J .$$

Therefore the projection on Ω of $\{g \circ X^- \neq f \circ X^-\}$ is contained in

$$\bigcup_n \{g \circ X_{T_n-} < f \circ X_{T_n-},\ T_n < \infty\} ,$$

and for each n

$$P^{\mu}(g \circ X_{T_n -} < f \circ X_{T_n -},\ T_n < \infty) = \nu_n(f-g) \le 2^n \nu(f-g) = 0,$$

completing the proof of (11.11-i).

If $f \in b\underline{\underline{E}}_r$, then by (11.3) there exists $\bar{f} \in b\overline{\underline{\underline{E}}}$ such that $f = 1_E \bar{f}$. Then $\bar{f} - f$ satisfies the hypotheses of (i) and $\bar{f} \circ X^-$ is previsible because $\bar{f} \in b\overline{\underline{\underline{E}}}$. Therefore (ii) follows from (i) and the proof of (11.11) is complete.

REMARK. It is immediate from (11.11) that if $A \in \overline{\underline{\underline{E}}}^*$ with $A \subset \bar{E} - E$, then the set $\{(t, \omega): X_{t-}(\omega) \in A\}$ is P^{μ} indistinguishable from a previsible set over $(\Omega, \underline{\underline{F}}^{\mu}, \underline{\underline{F}}_t^{\mu}, P^{\mu})$ for each μ on E.

It is now possible to obtain a bit more information about $X^- = (X_{t-})$ in the present case. For this we make the following definition.

(11.12) DEFINITION. *If μ is a probability on $(E, \underline{\underline{E}})$, a set $A \subset \bar{E}$ is μ-useless provided both of the sets $\{(t, \omega): X_t(\omega) \in A\}$ and $\{(t, \omega): X_{t-}(\omega) \in A\}$ are P^{μ} evanescent. A set $A \subset \bar{E}$ is useless if it is μ-useless for all such μ.*

(11.13) PROPOSITION. *Let $N = \{x \in \bar{E}: \bar{P}_0(x, \bar{E} - E) > 0\}$. Then N is useless.*

PROOF. Because $\bar{P}_0(x, \cdot) = \epsilon_x$ if $x \in D$ and $E \subset D$, it is clear that $D - E \subset N \subset \bar{E} - E$. Hence $\{(t, \omega): X_t(\omega) \in N\}$ is even empty. Since $\bar{E} - E$ is in $\overline{\underline{\underline{E}}}^*$, it is clear that N is in $\overline{\underline{\underline{E}}}^*$ also. Now fix μ on E and let all statements refer to the system $(\Omega, \underline{\underline{F}}^{\mu}, \underline{\underline{F}}_t^{\mu}, P^{\mu})$. Then by (11.11), $\Gamma = \{X^- \in N\}$ is (indistinguishable from) a previsible set. Thus if Γ is not P^{μ} evanescent there exists a previsible stopping time T with $[[T]] \subset \Gamma$ and $P^{\mu}[T < \infty] > 0$. Therefore using (5.11) and the fact that $T > 0$ because $[[T]] \subset \Gamma$ one obtains

$$0 = P^{\mu}[X_T \in \bar{E} - E,\ T < \infty] = E^{\mu}[\bar{P}_0(X_{T-}, \bar{E} - E),\ T < \infty].$$

This last formula states that $X_{T-} \in \{x: \bar{P}_0(x, \bar{E} - E) = 0\}$ almost surely on $\{T < \infty\}$ contradicting $[[T]] \subset \Gamma$. This establishes (11.13).

Proposition 11.13 has several important corollaries. Recall that D is the set of nonbranch points of $(\bar{P}_t)$ and that $E \subset D$.

(11.14) COROLLARY. *The set $D - E$ is useless.*

PROOF. If $x \in D - E$, then $\overline{P}_0(x, \cdot) = \varepsilon_x$ which is carried by $\overline{E} - E$. Therefore $D - E \subset N$ where N is the set defined in the statement of (11.13). Since a subset of a useless set is useless, this establishes (11.14).

Note that (11.14) states that almost surely if $X_{t-}(\omega)$ is not in E, then it must be a branch point of $(\overline{P}_t)$. In particular $\{X^- \notin E\}$ is indistinguishable from the previsible set $\{X^- \notin D\}$.

(11.15) COROLLARY. (i) Let $\overline{f} \in b\overline{\underline{E}}$ and $f = 1_E \overline{f}$. Then the processes $(\overline{P}_0 \overline{f} \circ X_{t-})$ and $(\overline{P}_0 f \circ X_{t-})$ are indistinguishable.

(ii) If $f \in b\underline{E}$, then for each μ on E the previsible projection of $f \circ X = (f \circ X_t)$ relative to $(\Omega, \underline{F}^\mu, \underline{F}^\mu_t, P^\mu)$ is P^μ indistinguishable from $\overline{P}_0 f \circ X^- = (\overline{P}_0 f \circ X_{t-})$. Recall that f is extended to $\overline{E}$ by setting $f = 0$ on $\overline{E} - E$.

PROOF. Since $\{x: \overline{P}_0 \overline{f}(x) \neq \overline{P}_0 f(x)\}$ is contained in $N = \{x: \overline{P}_0(x, \overline{E} - E) > 0\}$ (i) is an immediate consequence of (11.13). For (ii) fix μ on E and let all statements refer to the system $(\Omega, \underline{F}^\mu, \underline{F}^\mu_t, P^\mu)$. If $f \in b\underline{E}$, then by the last assertion of (11.3) there exists $\overline{f} \in b\overline{\underline{E}}$ such that $f = 1_E \overline{f}$. Thus by (i), $\overline{P}_0 f \circ X_{t-}$ is indistinguishable from the previsible process $\overline{P}_0 \overline{f} \circ X_{t-}$. On the other hand (5.11) implies that if T is previsible, then

$$E^\mu[f \circ X_T, T < \infty] = E^\mu[\overline{P}_0 f \circ X_{T-}, T < \infty],$$

proving (ii). (Observe that $\overline{P}_0 f(X_{0-}) = \overline{P}_0 f(X_0) = f(X_0)$ since $X_0 \in E$.)

Thus the machinery that we have developed enables us to compute explicitly the previsible projection of $f \circ X$ for each $f \in b\underline{E}$. Following Dellacherie we shall denote the previsible projection of a process Z by ${}^3Z$. We close this section with the following result of M. J. Sharpe. In its statement (g_j) is our fixed countable dense subset of $\underset{\sim}{R}$ and $\overline{g}_j$ is the unique continuous extension of g_j to $\overline{E}$.

(11.16) PROPOSITION. The sets $\Gamma_1 = \{X^- \notin E\}$ and $\Gamma_2 = \bigcup \{{}^3(g_n \circ X) \neq \overline{g}_n \circ X^-\}$ are indistinguishable.

PROOF. By (11.15-ii), Γ_2 is indistinguishable from

$$\Gamma_3 = \bigcup_n \{\overline{P}_0 g_n \circ X^- \neq \overline{g}_n \circ X^-\} .$$

If $x \in E$, $\overline{P}_0 g_n(x) = g_n(x) = \overline{g}_n(x)$. Therefore $\Gamma_3 \subset \Gamma_1$. By (11.11) and (11.15), Γ_1 and Γ_3 are previsible, and so if $\Gamma_1 - \Gamma_3$ is not evanescent there exists a μ on E and a previsible $(\underline{\underline{F}}_t^\mu)$ stopping time T such that

(11.17) $\quad P^\mu[X_{T-} \notin E, \overline{P}_0 g_n(X_{T-}) = \overline{g}_n(X_{T-})$ for all n, $T < \infty] > 0$.

By (11.15-i), $\overline{P}_0 g_n(X_{T-}) = \overline{P}_0 \overline{g}_n(X_{T-})$ almost surely P^μ on $\{T < \infty\}$, and so one may replace $\overline{P}_0 g_n$ by $\overline{P}_0 \overline{g}_n$ in (11.17). On the other hand if $\overline{P}_0 \overline{g}_n(x) = \overline{g}_n(x)$ for all n, then since $(\overline{g}_n)$ is dense in $\overline{\underset{\sim}{R}}$ and $\overline{\underset{\sim}{R}} - \overline{\underset{\sim}{R}}$ is dense in $\underset{\sim}{C}(\overline{E})$ it follows that $\overline{P}_0 \overline{f}(x) = \overline{f}(x)$ for all $f \in \underset{\sim}{C}(\overline{E})$. Consequently $x \in D$ according to (3.6-iv). Combining these remarks with (11.17) we have

$$P^\mu[X_{T-} \notin E, X_{T-} \in D, T < \infty] > 0,$$

and this contradicts the fact that $D - E$ is useless, proving (11.16).

12. RIGHT PROCESSES CONTINUED: SHIH'S THEOREM

The most striking application to date of the machinery developed in Sections 10 and 11 is the proof of Hunt's balayage theorem for right processes. The crucial step in extending Hunt's original proof to right processes is a theorem of Shih [15]. Before coming to these results we need to introduce a few of the standard definitions in the theory of Markov processes. In Hunt's original approach to probabilistic potential theory as expounded, for example, in BG the nearly Borel measurability of the excessive functions plays a key role. As mentioned previously the reason for this is that the process $f \circ X = (f \circ X_t)$ is well measurable when f is nearly Borel. However, we are not assuming that the excessive functions are nearly Borel and so we must exercise some care in our definitions. It will turn out that this added generality is, at best, an illusion since in the Ray topology the excessive functions will be "nearly Borel."

We fix a right process X and we shall use the notations and results of the preceding sections without special comment. However, we remind the reader that Ω is the set of all maps $\omega: \mathbb{R}^+ \to E$ that are right continuous in both the original and Ray topologies and which have left limits in $\overline{E}$ in the ρ-topology. We shall change our notation slightly for the various spaces of continuous functions. We shall write $\underset{\sim}{C}_u(d) = \underset{\sim}{C}_u(E, d)$, resp. $\underset{\sim}{C}_u(\rho) = \underset{\sim}{C}_u(E, \rho)$, for the space of bounded real valued d-uniformly, resp. ρ-uniformly, continuous functions on E. (We previously wrote $\underset{\sim}{C}_u$ for $\underset{\sim}{C}_u(d)$, but now the interplay between the two metrics d and ρ will be of importance and we need a consistent notation.) Of course, $\underset{\sim}{C}_u(d)$, resp. $\underset{\sim}{C}_u(\rho)$, is just the set of restrictions to E of the continous functions on $\hat{E}$, resp. $\overline{E}$. Also recall that $\underline{\underline{E}}$, resp. $\underline{\underline{E}}_r$, is the σ-algebra of Borel subsets of E in the original, resp. Ray, topology on E. According to (8.4-ii), $\underline{\underline{E}} = \sigma(\underset{\sim}{C}_u(d))$ and $\underline{\underline{E}}_r = \sigma(\underset{\sim}{C}_u(\rho))$, while by (11.3), $\underline{\underline{E}} \subset \underline{\underline{E}}_r = \sigma(\underset{\sim}{R})$ and $\underline{\underline{E}}^* = \underline{\underline{E}}_r^*$.

(12.1) LEMMA. For each $\alpha > 0$, U^{α} maps $\underset{\sim}{C}_u(\rho)$ into $\underset{\sim}{C}_u(\rho)$ and also $b\underset{=}{E}_r$ into $b\underset{=}{E}_r$. For each $t \geq 0$, P_t maps $b\underset{=}{E}_r$ into $b\underset{=}{E}_r$.

PROOF. If $f \in \underset{\sim}{C}_u(\rho)$, then $f = \bar{f}\big|_E$ where $\bar{f} \in \underset{\sim}{C}(\bar{E})$. By (11.9), $U^{\alpha}f(x) = \bar{U}^{\alpha}\bar{f}(x)$ if $x \in E$; that is, $U^{\alpha}f = \bar{U}^{\alpha}\bar{f}\big|_E$. But $\bar{U}^{\alpha}\bar{f} \in \underset{\sim}{C}(\bar{E})$ because $(\bar{U}^{\alpha})$ is a Ray resolvent, and so U^{α} maps $\underset{\sim}{C}_u(\rho)$ into $\underset{\sim}{C}_u(\rho)$. It is immediate from this and (8.6) applied to $\underset{\sim}{C}_u(0)$ and $\underset{=}{E}_r$ that U^{α} sends $b\underset{=}{E}_r$ into $b\underset{=}{E}_r$. If $f \in \underset{\sim}{C}_u(\rho)$, then $t \to P_t f(x) = E^x[f(X_t)]$ is right continuous. Since $U^{\alpha}f \in b\underset{=}{E}_r$ it follows from (3.12) that $P_t f \in b\underset{=}{E}_r$ whenever $f \in \underset{\sim}{C}_u(\rho)$. Using (8.6) again this establishes the second sentence in (12.1), completing the proof.

(12.2) DEFINITION. A universally measurable function f on E is well measurable provided that for each μ on E the process $f \circ X = (f \circ X_t)$ is P^{μ} indistinguishable from a well measurable process over the system $(\Omega, \underset{=}{F}^{\mu}, \underset{=}{F}_t^{\mu}, P^{\mu})$.

(12.3) DEFINITION. A numerical function f on E is nearly Ray Borel provided that for each μ on E there exist $g, h \in \underset{=}{E}_r$ with $g \leq f \leq h$ such that the processes $g \circ X$ and $h \circ X$ are P^{μ} indistinguishable.

Let $\underset{=}{W}$ (resp. $\underset{=}{E}_r^n$) be the class of all subsets A of E such that 1_A is well measurable (resp. nearly Ray Borel). It is easy to see that $\underset{=}{W}$ and $\underset{=}{E}_r^n$ are σ-algebras on E. The reader should check that a numerical function f on E is well measurable (resp. nearly Ray Borel) if and only if f is $\underset{=}{W}$ (resp. $\underset{=}{E}_r^n$) measurable. If $g \in \underset{=}{E}_r$, then $g = \bar{g}\big|_E$ with $\bar{g} \in \underset{=}{\bar{E}}$ and since $X_t(\omega) \in E$ for all t and ω, $g \circ X = \bar{g} \circ X$ is well measurable. Since $\underset{=}{E}_r^* = \underset{=}{E}^*$ by (11.3), this yields $\underset{=}{E}_r^n \subset \underset{=}{W} \subset \underset{=}{E}^*$. Axiom HD2 states that each α-excessive function is well measurable. The next lemma shows that we can do a bit better. Recall that $\mathcal{E}^{\alpha}$ denotes the class of α-excessive functions.

(12.4) LEMMA. (i) Each α-excessive function is nearly Ray Borel.
   (ii) Let $\underset{=}{B}^e(E) = \sigma(\cup_{\alpha} \mathcal{E}^{\alpha})$. Then $\underset{=}{E} \subset \underset{=}{E}_r \subset \underset{=}{B}^e(E) \subset \underset{=}{E}_r^n \subset \underset{=}{W} \subset \underset{=}{E}^*$.
   (iii) For each $t \geq 0$, P_t sends $b\underset{=}{B}^e(E)$ into itself.

PROOF. For (i) it suffices to show that $U^{\alpha}g$ is $\underset{=}{E}_r^n$ measurable whenever $\alpha > 0$ and $g \in b\underset{=}{E}_+^*$. But $\underset{=}{E}^* = \underset{=}{E}_r^*$, and so given μ on E if $\nu = \mu U^{\alpha}$ there exist $g_1, g_2 \in b\underset{=}{E}_r^+$ with $g_1 \leq g \leq g_2$ and $\nu(g_1) = \nu(g_2)$. Clearly $U^{\alpha}g_1 \leq U^{\alpha}g$

$\le U^\alpha g_2$ and according to (12.1), $U^\alpha g_1$ and $U^\alpha g_2$ are $\underline{\underline{E}}_r$ measurable. Now exactly as in the proof of (5.8)

$$E^\mu[U^\alpha g_2(X_t) - U^\alpha g_1(X_t)] \le e^{\alpha t}\ \nu(g_2 - g_1) = 0$$

for each fixed t. However $U^\alpha g_j \in \mathcal{E}^\alpha$ for $j = 1, 2$ and so it follows from HD2 that the processes $U^\alpha g_j \circ X$ for $j = 1, 2$ are P^μ indistinguishable. This establishes (12.4-i).

We already know that $\underline{\underline{E}} \subset \underline{\underline{E}}_r \subset \underline{\underline{E}}^n_r \subset \underline{\underline{W}} \subset \underline{\underline{E}}^*$ and $\underline{\underline{B}}^e(E) \subset \underline{\underline{E}}^n_r$ by (i). From (11.3), $\underline{\underline{E}}_r = \sigma(\underset{\sim}{R})$ and since each element of $\underset{\sim}{R}$ is β-excessive for some β, this implies $\underline{\underline{E}}_r \subset \underline{\underline{B}}^e(E)$, completing the proof of (12.4-ii).

For the proof of (iii) let $\mathcal{M} = b \cup \mathcal{E}^\alpha$ so $\sigma(\mathcal{M}) = \underline{\underline{B}}^e(E)$. Let f be a finite product of functions in $\mathcal{M}$. Then $t \to f(X_t)$ is almost surely right continuous and so $t \to P_t f(x)$ is right continuous. Clearly $U^\alpha f \in b\underline{\underline{E}}^e(E)$ for each $\alpha > 0$. Therefore by (3.12), $P_t f$ is $\underline{\underline{B}}^e(E)$ measurable whenever f is a finite product of functions in $\mathcal{M}$. Consequently by D-IV-T18, P_t maps $b\underline{\underline{B}}^e(E)$ into $b\underline{\underline{B}}^e(E)$, completing the proof of (12.4).

The content of (12.1) and (12.4) is that by changing the topology on E to the Ray topology the resolvent and semigroup become Borel and the excessive functions become nearly Borel. In some arguments it is handy to note that the σ-algebras $\underline{\underline{F}}^{0r}$, resp. $\underline{\underline{F}}^{0r}_t$, generated by the X_s, resp. X_s with $s \le t$, when considered as maps from Ω to the measurable space $(E, \underline{\underline{E}}_r)$ are separable and that $\underline{\underline{F}}^\mu$ is the P^μ completion of $\underline{\underline{F}}^{0r}$ while $\underline{\underline{F}}^\mu_t$ is generated by $\underline{\underline{F}}^{0r}_t$ and all P^μ null sets in $\underline{\underline{F}}^\mu$. In [12], Meyer has shown, at least when E is Lusinien, that if f is well measurable, then for each μ on E there exists $f' \in \underline{\underline{E}}$ such that $f' \circ X$ and $f \circ X$ are P^μ indistinguishable. Thus the differences between $\underline{\underline{E}}^n$, $\underline{\underline{E}}^n_r$, and $\underline{\underline{W}}$ are not very great. We shall have no need for Meyer's result and so shall not prove it here.

If B is any subset of $\overline{E}$ we define

(12.5) $$T_B = \inf\{t > 0 \colon X_t \in B\}$$

(12.6) $$D_B = \inf\{t \ge 0 \colon X_t \in B\}\ .$$

Note that T_B and D_B are unchanged if we replace B by $B \cap E$. If $B \in \underline{\underline{W}}$ and μ is a probability on E, then $\{(t, \omega) \colon t > 0,\ X_t(\omega) \in B\}$ is $(P^\mu$

indistinguishable from) a well measurable set over $(\Omega, \underline{\underline{F}}^{\mu}, \underline{\underline{F}}_t^{\mu}, P^{\mu})$ and T_B is its debut. Therefore by D-III-T23, T_B is an $(\underline{\underline{F}}_t^{\mu})$ stopping time, and since μ is arbitrary T_B is an $(\underline{\underline{F}}_t)$ stopping time. Similarly, D_B is an $(\underline{\underline{F}}_t)$ stopping time if $B \in \underline{\underline{W}}$. T_B is called the <u>hitting time</u> of B and D_B the <u>entry time</u> of B. If T is an $(\underline{\underline{F}}_t)$ stopping time and $\alpha \geq 0$, we define for $f \in b\underline{\underline{E}}^*$

$$(12.7) \qquad P_T^{\alpha} f(x) = E^x\{e^{-\alpha T} f(X_T);\ T < \infty\} .$$

It is clear that $P_T^{\alpha}: b\underline{\underline{E}}^* \to b\underline{\underline{E}}^*$. If $T = T_B$ with $B \in \underline{\underline{W}}$ we write P_B^{α} in place of $P_{T_B}^{\alpha}$. It is not difficult to check that if $f \in \mathcal{E}^{\alpha}$ and $B \in \underline{\underline{W}}$, then $P_B^{\alpha} f \in \mathcal{E}^{\alpha}$. See BG-II-(2.8).

The following simple but important fact is known as Blumenthal's zero-one law.

(12.8) PROPOSITION. <u>If</u> $x \in E$ <u>and</u> $A \in \underline{\underline{F}}_0^{\varepsilon_x}$, <u>then</u> $P^x(A)$ <u>is either zero or one</u>.

PROOF. Using the Markov property and $P^x(X_0 = x) = 1$ we have for such an A

$$P^x(A) = P^x(A \cap A) = E^x\{P^{X(0)}(A);\ A\} = P^x(A)^2 ,$$

proving (12.8).

Note that (12.8) implies that if $A \in \underline{\underline{F}}_0$, then for each $x \in E$, $P^x(A)$ is either zero or one. If T is an $(\underline{\underline{F}}_t)$ stopping time, then $\{T = 0\} \in \underline{\underline{F}}_0$. We say that x is regular for T if $P^x(T = 0) = 1$ and that x is irregular for T if $P^x(T > 0) = 1$. By (12.8) each x in E is either regular or irregular for T. In particular if $B \in \underline{\underline{W}}$, we say that x is <u>regular</u>, resp. irregular, for B if $P^x(T_B = 0) = 1$, resp. $P^x(T_B > 0) = 1$. If B^r denotes the set of regular points for B, then it is evident that $B^r = \{x: E^x(e^{-T_B}) = 1\}$. One easily checks that $x \to E^x(e^{-T_B})$ is 1-excessive and so $B^r \in \underline{\underline{B}}^e(E)$. A set $B \in \underline{\underline{W}}$ is finely open if each $x \in B$ is <u>irregular</u> for $E - B$; that is, $P^x(T_{E-B} > 0) = 1$ for each $x \in B$. In other words B is <u>finely open</u> if the process starting from a point in B remains in B for an initial interval of time almost surely. Since the process is right continuous in both the original and Ray topologies, it follows that if $G \subset E$ is open in either topology, then G is finely open. We refer the reader to BG-Sec. II-4 for additional properties of finely open sets.

We are now in a position to state Hunt's theorem. The reader should bear in mind that $\underline{\underline{E}}^n \subset \underline{\underline{E}}_r^n$ since $\underline{\underline{E}} \subset \underline{\underline{E}}_r$.

(12.9) THEOREM. (Hunt). Let $f \in \mathcal{E}^\alpha$ and $B \in \underline{\underline{E}}_r^n$. Let $\mathcal{U} = \mathcal{U}(f, B) = \{u \in \mathcal{E}^\alpha : u \geq f \text{ on } B\}$, and let f_B be the lower envelope of $\mathcal{U}$; that is, $f_B = \inf\{u: u \in \mathcal{U}\}$. Then $P_B^\alpha f \leq f_B$. If $\alpha > 0$, $P_B^\alpha f(x) = f_B(x)$ except possibly for x in $B - B^r$. This last statement remains true when $\alpha = 0$ provided there exists $h \in b\underline{\underline{E}}_+^*$ with Uh bounded and strictly positive.

Hunt's original proof can be repeated to prove this result once the following theorem is established. Hunt's proof is given in detail in BG-III-(6.12) and also in [9], T18 of Ch. XV. We refer the reader to these sources and will not repeat the proof here. Of course, one must replace nearly Borel by nearly Ray Borel throughout the argument in these references.

(12.10) THEOREM. (Shih) (i) Let $B \in \underline{\underline{E}}_r^n$ and let μ be a probability on $(E, \underline{\underline{E}})$. Then there exists a decreasing sequence (H_n) of Ray open subsets of E with each $H_n \supset B$ and $D_{H_n} \uparrow D_B$ almost surely P^μ.

(ii) If, in addition, $\mu(B - B^r) = 0$, then $T_{H_n} \uparrow T_B$ almost surely P^μ.

PROOF. There exists $A \in \underline{\underline{E}}_r$ with $B \subset A$ and such that $\{(t, \omega): X_t(\omega) \in A - B\}$ is P^μ evanescent by the definition of a nearly Ray Borel set. Consequently there is no loss of generality in assuming that $B \in \underline{\underline{E}}_r$ in the proof. We shall prove (i) first. This represents the main work as (ii) is an easy corollary of (i).

In the course of the proof we shall need the following lemma from [10] which we state and prove here in order not to interrupt the proof of (12.10). For its statement recall the definition of a Choquet capacity from D-I-D28.

(12.11) LEMMA. Let F be a locally compact space with a countable base. Let $\underline{\underline{O}}$ and $\underline{\underline{K}}$ denote the open and compact subsets of F respectively. Let $I: \underline{\underline{O}} \to [0, \infty]$ satisfy:

(i) I is increasing; that is, $G, H \in \underline{\underline{O}}$, $G \subset H$ imply $I(G) \leq I(H)$.

(ii) I is strongly subadditive on $\underline{\underline{O}}$; that is, $I(G \cup H) + I(G \cap H) \leq I(G) + I(H)$ for $G, H \in \underline{\underline{O}}$.

(iii) $I(G) = \sup\{I(H): H \in \underline{\underline{O}},\ \overline{H} \in \underline{\underline{K}},\ \overline{H} \subset G\}$.

(iv) $I(G) < \infty$ if $\overline{G} \in \underline{\underline{K}}$.

(Here $\overline{A}$ denotes the closure of A when A is a subset of F.) Then

$I^*(B) = \inf\{I(G): G \supset B, G \in \underline{\underline{O}}\}$ _defined for all subsets_ B _of_ F _is a Choquet capacity relative to the paving_ $\underline{\underline{K}}$ _on_ F.

PROOF. We shall reduce this to the usual existence theorem for right continuous capacities. For $K \in \underline{\underline{K}}$ define $J(K) = I^*(K)$. Clearly J is increasing and positive on $\underline{\underline{K}}$, and by (iv), $J(K) < \infty$ for $K \in \underline{\underline{K}}$. Given $K \in \underline{\underline{K}}$ and $\epsilon > 0$ choose $G \in \underline{\underline{O}}$, $G \supset K$, and $I(G) \le J(K) + \epsilon$. If $K \subset K' \subset G$, then $J(K) \le J(K') \le I(G) \le J(K) + \epsilon$ and so J is right continuous on $\underline{\underline{K}}$. Finally J is strongly subadditive on $\underline{\underline{K}}$. To see this given $K, L \in \underline{\underline{K}}$ and $\epsilon > 0$ choose $G, H \in \underline{\underline{O}}$ with $G \supset K$, $H \supset L$, $I(G) \le J(K) + \epsilon$, and $I(H) \le J(L) + \epsilon$. Then using (ii)

$$J(K \cup L) + J(K \cap L) \le I(G \cup H) + I(G \cap H) \le I(G) + I(H) \le J(K) + J(L) + 2\epsilon ,$$

and since $\epsilon > 0$ is arbitrary, J is strongly subadditive on $\underline{\underline{K}}$. Now by T27 page 46 and the remark on page 47 of [8] if we define $J^*(G) = \sup\{J(K): K \subset G, K \in \underline{\underline{K}}\}$ for $G \in \underline{\underline{O}}$ and $J^*(B) = \inf\{J^*(G): G \supset B, G \in \underline{\underline{O}}\}$ for arbitrary subsets B of F, then J^* is a Choquet capacity on F relative to $\underline{\underline{K}}$ that is right continuous on the Borel sets of F. Note that $\underline{\underline{K}}_\delta = \underline{\underline{K}}$ in the present case. Thus to complete the proof of (12.11), it suffices to show $J^* = I^*$ and this is evident if $J^*(G) = I(G)$ for all $G \in \underline{\underline{O}}$. To this end fix $G \in \underline{\underline{O}}$. If $K \subset G$, then $J(K) = I^*(K) \le I(G)$ and so $J^*(G) \le I(G)$. If $H \in \underline{\underline{O}}$ with $\overline{H} \in \underline{\underline{K}}$ and $\overline{H} \subset G$, then

$$I(H) = I^*(H) \le I^*(\overline{H}) = J(\overline{H}) \le J^*(G).$$

But by (iii) this yields $I(G) \le J^*(G)$, completing the proof of (12.11).

PROOF OF (12.10). If $B \in \overline{\underline{\underline{E}}}$, then $B \cap E$ is in $\underline{\underline{E}}_r$ and so $D_B = D_{B \cap E}$ is an $(\underline{\underline{F}}_t)$ stopping time. For each $B \in \overline{\underline{\underline{E}}}$, define $I(B) = E^\mu[\exp(-D_B)]$. If $A, B \in \overline{\underline{\underline{E}}}$ and $A \subset B$, then $I(A) \le I(B)$ since $D_A \ge D_B$. We claim that I is strongly subadditive on $\overline{\underline{\underline{E}}}$. To see this first note that $D_{A \cup B} = D_A \wedge D_B$ and $D_{A \cap B} \ge D_A \vee D_B$. Therefore

$$((D_A \wedge D_B, D_A)) \subset ((D_B, D_A \vee D_B)) \subset ((D_B, D_{A \cap B})) ,$$

and so

$$I(A \cup B) - I(A) = E^{\mu}\left[e^{-D_{A\cup B}} - e^{-D_A}\right] = E^{\mu}\int_{D_A \wedge D_B}^{D_A} e^{-t}\,dt \le E^{\mu}\int_{D_B}^{D_{A\cap B}} e^{-t}\,dt$$

$$= I(B) - I(A \cap B),$$

proving the strong subadditivity of I on $\overline{\underline{E}}$. Next observe that if (B_n) is an increasing sequence in $\overline{\underline{E}}$ and $B = \cup B_n$, then $D_{B_n} \downarrow D_B$ and so $I(B_n) \uparrow I(B)$.

Let $\underline{K}$ and $\underline{O}$ denote the ρ-compact and ρ-open subsets of $\overline{E}$ respectively. Then $\underline{K} \subset \overline{\underline{E}}$ and $\underline{O} \subset \overline{\underline{E}}$. If $A \subset \overline{E}$, let $\overline{A}$ denote the ρ-closure of A in $\overline{E}$. Note that if $G \in \underline{O}$ then there exists a sequence $(H_n) \subset \underline{O}$ with $\overline{H}_n \in \underline{K}$, $\overline{H}_n \subset G$ and $H_n \uparrow G$. Consequently the restriction of I to $\underline{O}$ satisfies the hypotheses of (12.11) relative to the compact metric space $\overline{E}$. Therefore $I^*(B) = \inf\{I(G)\colon G \supset B,\ G \in \underline{O}\}$ defines a Choquet ($\underline{K}$) capacity on $\overline{E}$, and so by the capacitability theorem, D-I-T31, one has since $\underline{K}_\delta = \underline{K}$

(12.12) $$\sup\{I^*(K)\colon K \subset B,\ K \in \underline{K}\} = I^*(B) = \inf\{I(G)\colon G \supset B,\ G \in \underline{O}\}$$

for all $B \in \overline{\underline{E}}$.

In order to apply (12.12) we first show that if $K \in \underline{K}$ and $K \subset D$ then $I^*(K) = I(K) = E^{\mu}[\exp(-D_K)]$. Recall that D is the set of nonbranch points of $(\overline{P}_t)$ and that $D \in \overline{\underline{E}}$. Fix $K \in \underline{K}$ with $K \subset D$ and let (G_n) be a decreasing sequence in $\underline{O}$ with $G_n \supset \overline{G}_{n+1} \supset K$ for each n and $\cap G_n = K$. Such a sequence (G_n) exists since $\overline{E}$ is a compact metric space, and it is easy to check that $I^*(K) = \lim I(G_n)$. Let $R_n = D_{G_n}$. Then $X_{R_n} \in \overline{G}_n$ if $R_n < \infty$. Also $R_n \uparrow R \le D_K$. Let $\Lambda = \{R_n < R \text{ for all } n,\ R < \infty\}$ and $f = 1_K$. Then by (5.11)

(12.13) $$E^{\mu}[f(X_R)\,1_{\{R<\infty\}} \mid \vee\, \underline{F}^{\mu}_{R_n}] = f(X_R)\,1_{\Lambda^c}\,1_{\{R<\infty\}} + \overline{P}_0 f(X_{R-})\,1_{\Lambda}\,.$$

On $\Lambda^c \cap \{R < \infty\}$, $R_n = R$ for large n and so $X_R \in \cap \overline{G}_n = K$, that is $f(X_R) = 1_K(X_R) = 1$ on $\Lambda^c \cap \{R < \infty\}$. On Λ, $X_{R-} = \rho\text{-}\lim X_{R_n} \in K$ and since $K \subset D$ and $\overline{P}_0 f = f$ on D one has $\overline{P}_0 f(X_{R-}) = f(X_{R-}) = 1$ on Λ. Therefore the right side of (12.13) is $1_{\{R<\infty\}}$, and taking expectations one finds (recall $f = 1_K$)

$$P^{\mu}(X_R \in K;\ R < \infty) = P^{\mu}(R < \infty)\,.$$

Hence $X_R \in K$ almost surely P^μ on $\{R < \infty\}$, and so $D_K \le R$ almost surely P^μ. Therefore $D_K = R = \lim D_{G_n}$ almost surely P^μ, and so $I^*(K) = \lim I(G_n) = E^\mu[\exp(-D_K)] = I(K)$.

Now let $B \in \overline{\underline{E}}$ and $B \subset D$. Then by (12.12) and the above evaluation of $I^*(K)$ for $K \subset B \subset D$, there exist an increasing sequence $(K_n) \subset K$ with each $K_n \subset B$ and a decreasing sequence $(G_n) \subset \underline{O}$ with each $G_n \supset B$ such that $\sup I(K_n) = I^*(B) = \inf I(G_n)$. Then $D_{G_n} \le D_B \le D_{K_n}$ for each n and $D_{G_n} \uparrow R \le D_B$ while $D_{K_n} \downarrow S \ge D_B$. But

$$E^\mu(e^{-S}) = \lim I(K_n) = \lim I(G_n) = E^\mu(e^{-R}) \quad ,$$

and so $R = D_B = S$ almost surely P^μ. Consequently $D_{G_n} \uparrow D_B$ almost surely P^μ.

Next suppose $B \in \underline{E}_r$. Then there exists $A \in \overline{\underline{E}}$ with $B = A \cap E$. Since $D \in \overline{\underline{E}}$ and $E \subset D$ we may assume that $A \subset D$. By what was proved above there exists a decreasing sequence $(G_n) \subset \underline{O}$ with $G_n \supset A \supset B$ such that $D_{G_n} \uparrow D_A$ almost surely P^μ. For each n, $H_n = G_n \cap E$ is a Ray open subset of E, and since $D_{G_n} = D_{H_n}$ and $D_A = D_B$ statement (i) of (12.10) is established.

For (ii) first note that $T_{H_n} = D_{H_n}$ since H_n is Ray open. Clearly $D_B \le T_B$. If $x \in E$ and $x \notin B$, then $P^x(D_B < T_B) = 0$ because $P^x(X_0 \in B) = 0$. If $x \in B^r$, then $P^x(T_B = 0) = 1$, and so $P^x(D_B < T_B) = 0$. Therefore if $\mu(B - B^r) = 0$ and μ is carried by E, then

$$P^\mu(D_B < T_B) = \int P^x(D_B < T_B)\, \mu(dx) = 0 \quad .$$

Thus $D_B = T_B$ almost surely P^μ and the proof of (12.10-ii) is complete.

REMARK. A key step in the proof of (12.10) was the fact that $I(G_n) \downarrow I(K)$ whenever (G_n) was a decreasing sequence in $\underline{O}$ with $G_n \supset \overline{G}_{n+1} \supset K$ and $\cap G_n = K$ provided $K \in \underline{K}$ was contained in D. If one could prove this for arbitrary $K \in \underline{K}$, then one could apply T27, page 46, of [8] directly to the restriction of I to $\underline{K}$. But we only have this fact for $K \subset D$, and this is what forces us to use (12.11).

The most important application of Shih's theorem is in proving Hunt's theorem for right processes. However it is useful in many other situations. Here is a very simple application. Recall the notation of (11.10).

(12.14) PROPOSITION. <u>For any</u> $A \subset \bar{E}$ <u>let</u> $R_A = \inf\{t \geq 0: X_t \in A$ <u>or</u> $X_{t-} \in A\}$. <u>If</u> $B \in \underset{=r}{E}^n$, <u>then</u> $R_B = D_B$ <u>almost surely</u>.

PROOF. Clearly $R_A \leq D_A$ for all $A \subset \bar{E}$ and if G is ρ-open in $\bar{E}$ then $R_G = D_G$. Now fix $B \in \underset{=r}{E}^n$ and μ on $(E, \underset{=}{E})$. Then by (12.10), or more precisely its proof, there exists a decreasing sequence (G_n) of ρ-open subsets of $\bar{E}$ with $G_n \supset B$ for each n and $D_{G_n} \uparrow D_B$ almost surely P^μ. But $D_{G_n} = R_{G_n} \leq R_B \leq D_B$ and so $P^\mu[R_B < D_B] = 0$. Since μ is arbitrary (12.14) is established.

In Hunt's original approach to probabilistic potential theory a key technical fact was that T_B could be approached by T_K with K compact and $K \subset B$ provided B was nearly Borel. This approach is employed in BG, for example. However, the section theorems have now largely replaced the use of this fact and so its importance has diminished. The following general result of Dellacherie (see [11]) yields this approximation by compact subsets for entry times. Note that its proof uses the section theorem.

(12.15) PROPOSITION. <u>Let</u> $(\Omega, \underset{=}{F}, \underset{=t}{F}, P)$ <u>be a system satisfying the usual</u> <u>hypotheses of the general theory, and let</u> E <u>be a universally measurable subset</u> <u>of a compact metric space</u> $\hat{E}$. <u>Let</u> (Z_t) <u>be a well measurable process over the</u> <u>given system with values in</u> E. <u>If</u> B <u>is a Borel subset of</u> E, <u>define</u> D_B <u>to be</u> <u>the debut of</u> $\{(t, \omega): Z_t(\omega) \in B\}$. <u>Then there exists an increasing sequence</u> (K_n) <u>of compact subsets of</u> E <u>with</u> $K_n \subset B$ <u>for each</u> n <u>and</u> $D_{K_n} \downarrow D_B$ <u>almost</u> <u>surely</u>.

PROOF. By D-III-T23, D_B is a stopping time whenever B is a Borel subset of E since $\Lambda = \{(t, \omega): Z_t(\omega) \in B\}$ is well measurable. Fix $\epsilon > 0$. Then $\Lambda_\epsilon = \Lambda \cap [[D_B, D_B + \epsilon]]$ is well measurable. Therefore by the section theorem D-IV-T10 there exists a stopping time T with $[[T]] \subset \Lambda_\epsilon$ and $P(T < \infty) \geq P[\pi(\Lambda_\epsilon)] - \epsilon$ where π is the projection of Λ_ϵ on Ω. In other words $Z_T \in B$, $D_B \leq T \leq D_B + \epsilon$ on $\{T < \infty\}$ and since $\pi(\Lambda_\epsilon) = \{D_B < \infty\}$, $P(D_B < \infty, T = \infty) \leq \epsilon$.

Let $\mu(dx) = P[Z_T \in dx, T < \infty]$ be the distribution of Z_T. Then μ is carried by B, and by (8.4) may be regarded as a measure on $(\hat{E}, \hat{\underline{\underline{E}}}^*)$. Since any finite measure on $(\hat{E}, \hat{\underline{\underline{E}}}^*)$ is regular there exists a compact $K \subset B$ with $\mu(B - K) < \epsilon$. (Since $K \subset B \subset E$ it is clear that K is compact in $\hat{E}$ if and only if K is compact in E.) Now $D_B \le D_K$ and $D_K \le T$ on $\{Z_T \in K, T < \infty\}$. Observe that

$$P(Z_T \in K, T < \infty) = \mu(K) = \mu(B) - \mu(B - K) \ge P(T < \infty) - \epsilon ,$$

and therefore $P(D_K \le T < \infty) \ge P(T < \infty) - \epsilon$. Also

$$\{D_K > D_B + \epsilon, T < \infty\} \subset \{D_K > T, T < \infty\},$$

$$\{D_K > D_B + \epsilon, T = \infty\} \subset \{D_B < \infty, T = \infty\} ,$$

and so $P(D_K > D_B + \epsilon) \le 2\epsilon$.

Apply this with $\epsilon_n = 2^{-n}$ to obtain a sequence (K_n) of compact subsets of B with $P(D_{K_n} \ge D_B + 2^{-n}) \le 2^{-n+1}$. Let $L_n = \bigcup_{j=1}^{n} K_j \subset B$. Then (L_n) is an increasing sequence of compact subsets of B and $D_B \le D_{L_n} \le D_{K_n}$. Therefore $P[D_{L_n} > D_B + 2^{-n}] \le 2^{-n+1}$ and using the Borel-Cantelli lemma this implies that $D_{L_n} \downarrow D_B$ almost surely completing the proof of (12.15).

REMARKS. The reader may find it instructive to decide where this proof breaks down if one works with $T_B = \inf\{t > 0: Z_t \in B\}$ rather than D_B. Of course (12.15) applies immediately to our right process X to yield the following statement. If B is nearly Borel in E (i.e. $B \in \underline{\underline{E}}^n$), then for each μ on E there exists an increasing sequence (K_n) of compact (in the original topology of E) subsets of B with $D_{K_n} \downarrow D_B$ almost surely P^μ. One may repeat the argument in BG-I-(10.19) to obtain an increasing sequence (K_n) of compact subsets of B such that $T_{K_n} \downarrow T_B$ almost surely P^μ. One may just as well apply (12.15) to X as an $(E, \underline{\underline{E}}_r)$ process and obtain the analogous statements for the Ray topology.

13. COMPARISON OF (X_{t-}) AND (X^*_{t-})

In this section we shall give some additional applications of Ray processes to right processes. We shall be particularly concerned with the relationship betweeen X_{t-} and X^*_{t-}. See (11.10) for notation. Let us emphasize again that the object of primary interest is the right process X.

In this section we fix a right process X with state space E and we shall use the notation and terminology of the preceding sections without special mention. We begin by translating (6.4), (7.6), (5.15) and (6.8) to the present situation. Taking into account the facts that B_d is empty (10.9) and that $D - E$ is useless (11.14), the following result obtains.

(13.1) THEOREM. *Let μ be a probability on E and T an $(\underline{\underline{F}}^{\mu}_t)$ stopping time.*

(i) *If T is previsible, then $\underline{\underline{F}}^{\mu}_{T-} = \underline{\underline{F}}^{\mu}_T$ if and only if $P^{\mu}[X_{T-} \notin E, T < \infty] = 0$.*

(ii) *If $X_T = X_{T-}$ almost surely P^{μ} on $\{T < \infty\}$, then T is previsible and $\underline{\underline{F}}^{\mu}_{T-} = \underline{\underline{F}}^{\mu}_T$.*

(iii) *The totally inaccessible part of T is T_A where $A = \{X_{T-} \in E, X_T \neq X_{T-}, T < \infty\}$.*

(iv) *If (T_n) is a sequence of $(\underline{\underline{F}}^{\mu}_t)$ stopping times increasing to T, then $\{X_T = X_{T-}\} \cap \Lambda = \{X_{T-} \in E\} \cap \Lambda$ almost surely P^{μ} where $\Lambda = \{T_n < T$ for all $n, T < \infty\}$.*

(v) *The set $\{(t, \omega): X_{t-}(\omega) \in \bar{E} - E\}$ is P^{μ} indistinguishable from a countable union of graphs of previsible $(\underline{\underline{F}}^{\mu}_t)$ stopping times.*

The reader should check that in examples (10.12-iv) and (10.12-v) there exist previsible stopping times T for which $\underline{\underline{F}}^{\mu}_T \neq \underline{\underline{F}}^{\mu}_{T-}$ for appropriate μ. The next result characterizes those right processes for which the family $(\underline{\underline{F}}^{\mu}_t)$ is quasi-left-continuous for each μ on E. Such processes might be called *special* right processes.

(13.2) PROPOSITION. Let μ be a probability on E. Then the following statements relative to the system $(\Omega, \underline{\underline{F}}^{\mu}, \underline{\underline{F}}_t^{\mu}, P^{\mu})$ are equivalent.

(i) $(\underline{\underline{F}}_t^{\mu})$ is quasi-left-continuous.

(ii) If (T_n) is an increasing sequence of stopping times with limit T, then $\underline{\underline{F}}_T^{\mu} = \vee \underline{\underline{F}}_{T_n}^{\mu}$.

(iii) If (T_n) is an increasing sequence of stopping times with limit T and $f \in \underset{\sim}{R}$, then $f(X_T)\, 1_{\{T < \infty\}}$ is $\vee \underline{\underline{F}}_{T_n}^{\mu}$ measurable.

(iv) If (T_n) is an increasing sequence of stopping times with limit T, then $X_T = \rho\text{-}\lim X_{T_n}$ almost surely P^{μ} on $\{T < \infty\}$.

(v) $\bar{E} - E$ is μ-useless.

PROOF. The equivalence of (i) and (ii) is D-III-T51. It is evident that (ii) implies (iii). We next shall show that (iii) implies (i). To this end let T be previsible and let (T_n) announce T. Then $\underline{\underline{F}}_{T-}^{\mu} = \vee \underline{\underline{F}}_{T_n}^{\mu}$ by D-III-T35. Now $\bar{\underset{\sim}{R}} - \bar{\underset{\sim}{R}}$ is dense in $\underset{\sim}{C}(\bar{E})$ and if $\bar{f} \in \bar{\underset{\sim}{R}}$ and $f = \bar{f}|_E$, then $f \in \underset{\sim}{R}$. Hence, since $X_T \in E$ if $T < \infty$, (iii) implies that $\bar{f}(X_T)\, 1_{\{T<\infty\}}$ is $\underline{\underline{F}}_{T-}^{\mu}$ measurable for all $\bar{f} \in \underset{\sim}{C}(\bar{E})$. Therefore if $Z_t = W_t \bar{f}(X_t)$ with W a bounded previsible process and $\bar{f} \in \underset{\sim}{C}(\bar{E})$, then $Z_T\, 1_{\{T<\infty\}}$ is $\underline{\underline{F}}_{T-}^{\mu}$ measurable by D-IV-T21 and the preceding discussion. It now follows from (7.16) and the monotone class theorem that $Z_T\, 1_{\{T<\infty\}}$ is $\underline{\underline{F}}_{T-}^{\mu}$ measurable whenever Z is a bounded well measurable process. By the remark following the proof of D-IV-T21 this gives $\underline{\underline{F}}_{T-}^{\mu} = \underline{\underline{F}}_{T}^{\mu}$. Thus (iii) implies (i), and so (i), (ii), and (iii) are equivalent.

We shall complete the proof by showing (i) $\Rightarrow$ (v) $\Rightarrow$ (iv) $\Rightarrow$ (iii). If $f \in \underset{\sim}{R}$, then f is Ray continuous and so it is clear that (iv) implies (iii). Next we show that (v) $\Rightarrow$ (iv). Let (T_n) be an increasing sequence of stopping times with limit T. Let $\Lambda = \{T_n < T \text{ for all } n,\ T < \infty\}$. Then to prove (iv) it suffices to show that $X_T = \rho\text{-}\lim X_{T_n}$ almost surely P^{μ} on Λ. On Λ, $\rho\text{-}\lim X_{T_n} = X_{T-}$ and by (13.1-iv), $\{X_T = X_{T-}\} = \{X_{T-} \in E\}$ almost surely P^{μ} on Λ. But by (v), $P^{\mu}[X_{T-} \in \bar{E} - E,\ T < \infty] = 0$, and so $X_T = X_{T-}$ almost surely P^{μ} on Λ proving (iv). Finally it remains to show that (i) $\Rightarrow$ (v). If T is a previsible $(\underline{\underline{F}}_t^{\mu})$ stopping time, then $\underline{\underline{F}}_{T-}^{\mu} = \underline{\underline{F}}_{T}^{\mu}$ by (i). Hence (13.1-i) implies that $P^{\mu}[X_{T-} \notin E,\ T < \infty] = 0$. Combining this with (13.1-v) we see that $\{(t, \omega): X_{t-}(\omega) \in \bar{E} - E\}$ is P^{μ} evanescent; that is $\bar{E} - E$ is μ-useless. This completes the proof of (13.2).

(13.3) REMARK. Property (13.2-iv) is the quasi-left-continuity of the process X in the Ray topology. This is the essential ingredient for X to be a Hunt process in the Ray topology. According to (12.1), P_t is always a kernel on $(E, \underline{\underline{E}}_r)$ and so if E is Borel in $\bar{E}$ and $\bar{E} - E$ is useless, then X is a true Hunt process in the Ray topology (9.8). If (P_t) is a Borel right semigroup, then E is Borel in $\bar{E}$ by (11.3-i). Thus, aside from the fact that E need not be Borel in $\bar{E}$, a "special" right process is a Hunt process in the Ray topology. This helps to explain the role of special standard processes in earlier work.

We turn next to the relationship between X_{t-} and X^*_{t-}. See (11.10) for notation.

(13.4) PROPOSITION. Let

$$\Gamma = \{(t, \omega): X^*_{t-}(\omega) \text{ does not exist or } X^*_{t-}(\omega) \neq X_{t-}(\omega)\} .$$

Then for each probability μ on E, Γ is P^μ indistinguishable from a countable union of graphs of previsible $(\underline{\underline{F}}^\mu_t)$ stopping times. In addition, P^μ almost surely, for all t if $X_{t-} \in E$ and $X_{t-} \neq X_t$, then X^*_{t-} exists and $X^*_{t-} = X_{t-}$.

PROOF. Fix μ on E and let all statements be relative to the system $(\Omega, \underline{\underline{F}}^\mu, \underline{\underline{F}}^\mu_t, P^\mu)$. Recall that $\hat{E}$ is a compact metric space in which E (with its original topology) is a universally measurable subset. For the purpose of this proof let $\hat{X}_{t-}$ denote the left limit at $t > 0$ of $s \to X_s$ whenever it exists in $\hat{E}$ and $\hat{X}_{0-} = X_0$. Then X^*_{t-} exists if and only if $\hat{X}_{t-}$ exists and $\hat{X}_{t-} \in E$. Let

$$\Gamma_1 = \{(t, \omega): \hat{X}_{t-}(\omega) \text{ exists}\}$$

$$\Gamma_2 = \{(t, \omega): \hat{X}_{t-}(\omega) \text{ exists and } \hat{X}_{t-}(\omega) \notin E\}$$

$$\Gamma_3 = \{(t, \omega): X^*_{t-}(\omega) \text{ does not exist}\} = \Gamma_1^c \cup \Gamma_2 .$$

We shall write $\hat{X}^-$ for the map from Γ_1 to $\hat{E}$ defined by $\hat{X}^-: (t, \omega) \to \hat{X}_{t-}(\omega)$. Using this notation we have $\Gamma_2 = \Gamma_1 \cap \{\hat{X}^- \notin E\}$.

Let $(\hat{f}_n)$ be a sequence that is uniformly dense in $\underset{\sim}{C}(\hat{E})$. According to D-VI-T3 the processes $\bar{Z}^n$ and $\underline{Z}^n$ defined by $\bar{Z}^n_0 = \underline{Z}^n_0 = \hat{f}_n \circ X_0$, and for $t > 0$, by

$$\underline{Z}^n_t = \liminf_{s \uparrow t,\, s < t} \hat{f}_n \circ X_s \; ; \quad \bar{Z}^n_t = \limsup_{s \uparrow t,\, s < t} \hat{f}_n \circ X_s$$

are previsible for each n. It is evident that

$$\Gamma_1 = \bigcap_n \{(t, \omega): \underline{Z}_t^n(\omega) = \bar{Z}_t^n(\omega)\},$$

and hence Γ_1 is previsible. Also $1_{\Gamma_1} \hat{f}_n \circ \hat{X}^- = 1_{\Gamma_1} \bar{Z}^n$ is previsible for each n. Consequently $1_{\Gamma_1} \hat{f} \circ \hat{X}^-$ is previsible for each $\hat{f} \in \underset{\sim}{C}(\hat{E})$, and then by the monotone class theorem, for each $\hat{f} \in b\hat{\underline{\underline{E}}}$.

Since $\hat{f} \circ X$ is well measurable for each $\hat{f} \in b\hat{\underline{\underline{E}}}$, it follows that $1_{\Gamma_1} \hat{F}(\hat{X}^-, X)$ is well measurable for each $\hat{F} \in b(\hat{\underline{\underline{E}}} \otimes \hat{\underline{\underline{E}}})$. Taking $\hat{F}$ to be the indicator of the diagonal in $\hat{E} \times \hat{E}$ we see that $\Lambda = \{(t, \omega): \hat{X}_{t-}(\omega)$ exists and $\hat{X}_{t-}(\omega) \neq X_t(\omega)\}$ is well measurable. But for each ω the right continuous function $t \to X_t(\omega)$ can have at most a countable number of discontinuities, and so each ω section of Λ is countable. Therefore by D-VI-T33, $\Lambda = \bigcup [[R_n]]$ where each R_n is a stopping time. Since $X_t(\omega) \in E$, it is clear that

$$\Gamma_2 \subset \Lambda = \bigcup [[R_n]] .$$

We are now going to use a familiar argument to show that Γ_2 is previsible. (See the proofs of (6.9) and (11.11).) Since $[[R_n]] \subset \Lambda$, $\hat{X}_{R_n -}$ exists on $\{R_n < \infty\}$. For each n let

$$\nu_n(dx) = P^\mu[\hat{X}_{R_n -} \in dx, R_n < \infty]$$

and $\nu = \sum 2^{-n} \nu_n$. Then ν is a measure on $(\hat{E}, \hat{\underline{\underline{E}}}^*)$. Since $E \in \hat{\underline{\underline{E}}}^*$ there exists $A \in \hat{\underline{\underline{E}}}$ with $E \subset A$ and $\nu(A - E) = 0$. Now

$$\Gamma_2 = (\Gamma_1 \cap \{\hat{X}^- \notin A\}) \cup (\Gamma_1 \cap \{\hat{X}^- \in A - E\}. \tag{13.5}$$

By the result in the second paragraph of this proof the first set on the right side of (13.5) is previsible because $A \in \hat{\underline{\underline{E}}}$, while the projection on Ω of the second set on the right side of (13.5) is contained in

$$\bigcup_n \{\hat{X}_{R_n -} \in A - E, R_n < \infty\}.$$

But $P^\mu[\hat{X}_{R_n -} \in A - E, R_n < \infty] = \nu_n(A - E) = 0$ for each n by the choice of A. Consequently Γ_2 is previsible (more precisely, P^μ indistinguishable from a previsible set), and so $\Gamma_3 = \Gamma_1^c \cup \Gamma_2$ is previsible.

Now the set Γ that we are interested in may be written $\Gamma = \Gamma_3 \cup \Gamma_4 \cup \Gamma_5$ where

(13.6) $\Gamma_4 = \Gamma_1 \cap \{\hat{X}^- \in E, X^- \notin E\}$; $\Gamma_5 = \Gamma_1 \cap \{\hat{X}^- \in E, X^- \in E; \hat{X}^- \neq X^-\}$.

Here $X^- = (X_{t-})$. Thus to prove that Γ is previsible it suffices to show that Γ_4 and Γ_5 are previsible. We have seen above that if $\hat{f} \in b\hat{\underline{E}}$, then $1_{\Gamma_1} \hat{f} \circ \hat{X}^-$ is previsible, and hence so is $1_{\Gamma_1}(1_{\hat{E}-E}\hat{f}) \circ \hat{X}^- = 1_{\Gamma_2} 1_{\Gamma_1} \hat{f} \circ \hat{X}^-$. Therefore $1_{\Gamma_1}(1_E \hat{f}) \circ \hat{X}^-$ is previsible for all $\hat{f} \in b\hat{\underline{E}}$. Taking $\hat{f} = 1$ and using (11.11-i), or the remark following the proof of (11.11), it is evident that Γ_4 is previsible. Moreover, by (11.11-ii), $g \circ X^-$ is previsible for $g \in b\underline{E} \subset b\underline{E}_r$. Since $\underline{E}$ is the trace of $\hat{\underline{E}}$ on E, this implies that $1_{\Gamma_1} F(\hat{X}^-, X^-)$ is previsible for each $F \in b(\underline{E} \otimes \underline{E})$ where it is understood that F is extended to $\hat{E} \times \bar{E}$ by setting $F = 0$ off $E \times E$. Let

$$G(x, y) = \sum_n (2^n \|f_n\|)^{-1} |f_n(x) - f_n(y)|$$

where (f_n) is a countable dense subset of $\underset{\sim}{C}_u(d)$. It is immediate that $G \in b(\underline{E} \otimes \underline{E})$ and since $\underset{\sim}{C}_u(d)$ separates the points of E,

$$\Gamma_5 = \Gamma_1 \cap \{1_{\Gamma_1} G(\hat{X}^-, X^-) > 0\}$$

is previsible. Therefore Γ is previsible.

Next observe that if $(t, \omega) \in \Gamma$, then t must be a point of discontinuity of $s \to X_s(\omega)$ either in the original topology or in the Ray topology. Since a right continuous map from $\mathbb{R}^+$ to a metric space can have at most a countable number of discontinuities, we see that each ω section of Γ is countable. It now follows from D-VI-T33 that Γ is (indistinguishable from) a countable union of graphs of previsible stopping times, proving the first assertion in (13.4).

For the second let $A = \{X^- \in E, X^- \neq X\}$. By (13.1-iii), $A \cap [[T]]$ is evanescent whenever T is a previsible stopping time. Consequently $\Gamma \cap A$ is evanescent, and this is precisely the second assertion in (13.4).

REMARK. The second assertion in (13.4) may be restated as follows. Each discontinuity t of $s \to X_s$ in the Ray topology at which $X_{t-} \in E$ is also a simple jump discontinuity of $s \to X_s$ in the original topology and the "jumps" are the same in the two topologies.

Proposition (13.4) gives rise to a criterion that a stopping time be accessible. This should be compared with (13.1-ii) and (13.1-iii).

(13.7) COROLLARY. *Let μ be a probability on E and T an $(\underline{\underline{F}}_t^\mu)$ stopping time. If P^μ almost surely on $\{0 < T < \infty\}$ either X^*_{T-} does not exist or $X^*_{T-} = X_T$, then T is accessible.*

PROOF. Let R be a totally inaccessible $(\underline{\underline{F}}_t^\mu)$ stopping time and let Γ be the set in the statement of (13.4). Since $[[R]] \cap \Gamma$ is P^μ evanescent, X^*_{R-} exists and equals X_{R-} almost surely P^μ on $\{R < \infty\}$, and by (13.1-iii), $X_{R-} \neq X_R$ almost surely P^μ on $\{R < \infty\}$. (Recall that $P^\mu(R = 0) = 0$ because R is totally inaccessible.) Therefore $P^\mu(R = T < \infty) = 0$ and (13.7) follows from D-III-T42.

We are going to study the process (X_{t-}) when the original process is a Hunt process, a standard process, or a special standard process. See (9.8), (9.9), and (9.10) for the definitions. It is easy to see that if X is a Hunt (resp. standard) process, then almost surely X^*_{t-} exists on $(0, \infty)$ (resp. $(0, \zeta)$), since Γ_3 in the proof of (13.4) is a countable union of graphs of previsible stopping times.

(13.8) PROPOSITION. (i) *If X is a Hunt process, then (X^*_{t-}) and (X_{t-}) are indistinguishable.*

(ii) *If X is a standard process, then almost surely if $t < \zeta$, either $X^*_{t-} = X_{t-}$ or $X_{t-} \notin E$ and $X^*_{t-} = X_t$.*

(iii) *If X is a special standard process, then almost surely if $t < \zeta$, $X^*_{t-} = X_{t-}$.*

PROOF. Fix a probability μ on E and all statements will refer to the system $(\Omega, \underline{\underline{F}}^\mu, \underline{\underline{F}}_t^\mu, P^\mu)$. Let X be a Hunt process. Then $(\underline{\underline{F}}_t^\mu)$ is quasi-left-continuous and so by (13.1-i) if T is previsible, $X_{T-} \in E$ almost surely on $\{T < \infty\}$. Since a Hunt process is a Borel right process, $E \in \overline{\underline{\underline{E}}}$ by (11.3-i). Therefore

$$H = \{(t, \omega) \colon 0 < t < \infty,\ X^*_{t-}(\omega) \neq X_{t-}(\omega)\}$$

is previsible. (This is easy in this case, but also note that $H = \Gamma_4 \cup \Gamma_5$ where

Γ_4 and Γ_5 are defined in (13.6) and hence H is previsible.) Thus if H is not evanescent there exists a previsible stopping time T with $[[T]] \subset H$ and $P^{\mu}(T < \infty) > 0$. But by (13.2-iv), $X_{T-} = X_T$ almost surely on $\{T < \infty\}$, while the quasi-left-continuity of X in the original topology yields $X^*_{T-} = X_T$ almost surely on $\{T < \infty\}$. This contradicts $[[T]] \subset H$ and so H is evanescent, establishing (i).

We next prove (ii). Let

$$J = \{(t, \omega): 0 < t < \zeta(\omega),\ X^*_{t-}(\omega) \neq X_{t-}(\omega),\ X_{t-}(\omega) \in E\} .$$

Then $J \subset \Gamma$ where Γ is the set appearing in the statement of (13.4). If T is a previsible stopping time, then by (13.1-iv), $X_T = X_{T-}$ almost surely on $\{X_{T-} \in E,\ T < \infty\}$. Also since X is standard $X^*_{T-} = X_T$ almost surely on $\{T < \zeta\}$. Combining these remarks with (13.4) we see that J is evanescent. In other words almost surely

$$\{X^*_{t-} \neq X_{t-},\ t < \zeta\} \subset \{X_{t-} \notin E\} .$$

On the other hand using (13.1-v) and the fact that X is standard

$$\{X_{t-} \notin E,\ t < \zeta\} \subset \{X^*_{t-} = X_t\}$$

almost surely. Combining these inclusions yields $\{X^*_{t-} \neq X_{t-};\ t < \zeta\} \subset \{X^*_{t-} = X_t,\ X_{t-} \notin E\}$ almost surely, proving (ii).

If X is special standard, then $(\underline{\underline{F}}^{\mu}_t)$ is quasi-left-continuous. Therefore (iii) is an immediate consequence of (ii) and (13.2-v).

The examples in (10.12) should be examined in light of the results of this section. The process in (10.12-v) is the simplest example of a standard process that is not special standard. If $\mu = \varepsilon_{-1}$, then for the process in (10.12-v),

$$P^{\mu}[X_{1-} \notin E,\ X^*_{1-} = X_1,\ 1 < \zeta\} = 1/2 ,$$

and so the second alternative in (13.8-ii) is indeed possible. Note that $T_{\{0\}}$ is accessible but not previsible under P^{μ}. Thus the hypotheses of (13.7) do not imply that T is previsible even for standard processes. In [16], Meyer and Walsh have shown that the behavior exhibited in this example is "typical" for standard processes. This is an interesting result and we refer the reader to [16] for it.

In example (10.12-iv) with $\mu = \epsilon_{-1}$, almost surely P^{μ}, $X_1 \neq X_{1-}$ and X^*_{1-} does not exist. Thus in the second assertion of (13.4), the condition $X_{t-} \in E$ is crucial. Since $\bar{E} - E$ is not μ-useless, $(\underline{\underline{F}}^{\mu}_t)$ is not quasi-left-continuous. In this example both $T_{\{1+\}}$ and $T_{\{1-\}}$ are accessible but not previsible under P^{μ}. This example is not a standard process although it is "essentially" the same example as (10.12-v).

14. U-SPACES

In this section we shall develop some additional properties of U-spaces that will be needed in Section 15. We refer the reader to the first part of Section 8 for the definition and elementary properties of U-spaces. In particular we recall the notation that if (E, d) is a metric space, then $\underset{\sim}{C}_b(E)$ denotes the real valued bounded continuous functions on E and $\underset{\sim}{C}_u(E, d) = \underset{\sim}{C}_u(d)$ denotes the real valued bounded d-uniformly continuous functions on E.

We next recall some facts about metric spaces and their completions. Let (E, d) be a metric space. Then there exists a complete metric space (M, ρ) and an isometry $i: E \to M$ such that $i(E)$ is dense in M. The complete metric space (M, ρ) is unique up to isometry and is called the completion of (E, d). The isometry i is called the injection of E into M. We shall often identify E with the subspace $i(E)$ of M. Let E_1 and E_2 be metric spaces with completions M_1 and M_2. If $\varphi: E_1 \to E_2$ is uniformly continuous (resp. an isometry), then there exists a unique uniformly continuous map (resp. isometry) $\psi: M_1 \to M_2$ such that $i_2 \circ \varphi = \psi \circ i_1$ where i_1 and i_2 are the injections of E_1 and E_2 into M_1 and M_2 respectively. It is well known that (M, ρ) is compact if and only if (E, d) is totally bounded.

Let E be a non-void set and let d_1 and d_2 be metrics on E. Let E_1 and E_2 denote the completions of (E, d_1) and (E, d_2) respectively. If $d_1 \leq d_2$, then the identity map e on E is uniformly continuous from (E, d_2) to (E, d_1) and so there exists a unique uniformly continuous map $\varphi: E_2 \to E_1$ such that $\varphi \circ i_2 = i_1 \circ e = i_1$ where i_1 and i_2 are the injections of (E, d_1) and (E, d_2) into E_1 and E_2 respectively. The following lemma collects some facts that will be used repeatedly in the sequel.

(14.1) LEMMA. *Let d_1 and d_2 be metrics on E with $d_1 \leq d_2$. Using the above notation:*

(i) *If E_2 is compact, then φ is a surjection of E_2 on E_1 and E_1 is compact.*

(ii) *If* d_1 *and* d_2 *induce the same topology on* E, *then* $i_2(E) = \varphi^{-1}[i_1(E)]$.

(iii) *If* E_2 *is compact and if* $i_2(E) = \varphi^{-1}[i_1(E)]$, *then* d_1 *and* d_2 *induce the same topology on* E.

PROOF. If E_2 is compact, then $\varphi(E_2)$ is compact, and hence closed, in E_1. But $\varphi(E_2) \supset \varphi \circ i_2(E) = i_1(E)$ which is dense in E_1. Therefore $\varphi(E_2) = E_1$, proving (i).

For (ii) we shall first prove the following statement:

(14.2) *Let* d_1 *and* d_2 *induce the same topology on* E. *If* $z \in E_2$ *and* $x \in E$ *with* $\varphi(z) = \varphi \circ i_2(x)$, *then* $z = i_2(x)$.

To establish this choose $(x_n) \subset E$ with $i_2(x_n) \to z$ in E_2. Since φ is continuous,

$$i_1(x_n) = \varphi \circ i_2(x_n) \to \varphi(z) = \varphi \circ i_2(x) = i_1(x) .$$

Therefore $x_n \to x$ in (E, d_1), and since the topologies are the same, $x_n \to x$ in (E, d_2), or equivalently $i_2(x_n) \to i_2(x)$ in E_2. Consequently $z = i_2(x)$, proving (14.2).

To establish (14.1-ii) we observe that $\varphi \circ i_2(E) = i_1(E)$ implies $i_2(E) \subset \varphi^{-1}[i_1(E)]$. Conversely if $z \in \varphi^{-1}[i_1(E)]$, then there exists an $x \in E$ with $\varphi(z) = i_1(x) = \varphi \circ i_2(x)$. By (14.2), $z = i_2(x)$ and so $z \in i_2(E)$, proving (14.1-ii).

Coming to (iii), we first note that by (i), E_1 is compact and φ is a surjection. Since $d_1 \le d_2$, if $(x_n) \subset E$ and $x_n \to x$ in (E, d_2), then $x_n \to x$ in (E, d_1). Therefore it suffices to show that if $(x_n) \subset E$ and $x_n \to x$ in (E, d_1), then $x_n \to x$ in (E, d_2). But this will follow if we show that each subsequence (x'_n) of (x_n) has a further subsequence (x''_n) with $x''_n \to x$ in (E, d_2), or equivalently, $i_2(x''_n) \to i_2(x)$ in E_2. Changing notation it suffices to show that if $x_n \to x$ in (E, d_1), then (x_n) has a subsequence (x'_n) with $i_2(x'_n) \to i_2(x)$ in E_2. Since E_2 is compact (x_n) has a subsequence (x'_n) with $i_2(x'_n) \to z \in E_2$. Now $i_1(x'_n) \to i_1(x)$ by assumption while $i_1(x'_n) = \varphi \circ i_2(x'_n) \to \varphi(z)$. Therefore $i_1(x) = \varphi(z)$, or $z \in \varphi^{-1}[i_1(E)]$. Thus by hypothesis there exists $y \in E$ with $z = i_2(y)$. Consequently $i_1(y) = \varphi \circ i_2(y) = \varphi(z) = i_1(x)$ and so $x = y$. Hence $i_2(x'_n) \to z = i_2(x)$, proving (iii).

We come now to the key fact that is needed for characterizing U-spaces.

(14.3) PROPOSITION. Let d_1 and d_2 be metrics on a set E and let E_1 and E_2 be the completions of (E, d_1) and (E, d_2). If d_1 and d_2 induce the same second countable topology on E, then $i_1(E)$ is universally measurable in E_1 if and only if $i_2(E)$ is universally measurable in E_2.

PROOF. Let $d = d_1 + d_2$. It is immediate that if d_1 and d_2 induce the same topology τ on E, then the topology induced by d on E is also τ. Thus it suffices to prove (14.3) when $d_1 \le d_2$. Hence in the remainder of this proof we assume that $d_1 \le d_2$, and as in the notation above (14.1), φ is the uniformly continuous map from E_2 to E_1 such that $\varphi \circ i_2 = i_1$.

We shall first show that if $i_1(E)$ is universally measurable in E_1, then $i_2(E)$ is universally measurable in E_2. This part of the argument does not use the assumption that the metric spaces (E, d_1) and (E, d_2) are second countable. Let μ be a finite measure on $(E_2, \underline{\underline{E}}_2)$ and let $\nu = \varphi(\mu)$ be its image under φ on $(E_1, \underline{\underline{E}}_1)$. Of course, φ, being continuous, is a measurable map from $(E_2, \underline{\underline{E}}_2)$ to $(E_1, \underline{\underline{E}}_1)$. Since $i_1(E) \in \underline{\underline{E}}_1^*$, there exist sets $A, B \in \underline{\underline{E}}_1$ with $A \subset i_1(E) \subset B$ and $\nu(B) = \nu(A)$. Using (14.1-ii) this yields

$$\varphi^{-1}(A) \subset \varphi^{-1}[i_1(E)] = i_2(E) \subset \varphi^{-1}(B)$$

and since $\varphi^{-1}(A), \varphi^{-1}(B) \in \underline{\underline{E}}_2$ and $\mu[\varphi^{-1}(A)] = \nu(A) = \nu(B) = \mu[\varphi^{-1}(B)]$, it follows that $i_2(E) \in \underline{\underline{E}}_2^*$.

Next assume $i_2(E) \in \underline{\underline{E}}_2^*$. Then we must show that $i_1(E) = \varphi[i_2(E)] \in \underline{\underline{E}}_1^*$. In light of (14.2) this is an immediate consequence of the following lemma.

(14.4) LEMMA. Let E and F be complete separable metric spaces and let φ be a measurable map from $(E, \underline{\underline{E}})$ to $(F, \underline{\underline{F}})$. Suppose that $A \in \underline{\underline{E}}^*$ has the property that if $x \in A$ and $z \in E$ with $\varphi(x) = \varphi(z)$, then $x = z$. Then $\varphi(A) \in \underline{\underline{F}}^*$.

PROOF. The hypothesis implies that φ restricted to A is injective. Let $B = \varphi(A)$ and let $\psi = \varphi^{-1}: B \to A$. Thus ψ is a bijection of B on A. Let $\underline{\underline{A}}$ be the trace of $\underline{\underline{E}}$ on A and $\underline{\underline{B}}^*$ the trace of $\underline{\underline{F}}^*$ on B. We assert that ψ is a measurable map from $(B, \underline{\underline{B}}^*)$ to $(A, \underline{\underline{A}})$. This amounts to showing that if $A_1 \in \underline{\underline{A}}$, then $\psi^{-1}(A_1) = \varphi(A_1) \in \underline{\underline{B}}^*$. Now $A_1 = A_0 \cap A$ with $A_0 \in \underline{\underline{E}}$, and $\varphi(A_1) \subset \varphi(A_0) \cap \varphi(A)$. However, if $y \in \varphi(A_0) \cap \varphi(A)$, then there exist $z \in A_0$

and $x \in A$ with $y = \varphi(z) = \varphi(x)$. Thus by hypothesis $x = z$, and so $y \in \varphi(A_0 \cap A)$. As a result $\varphi(A_1) = \varphi(A_0) \cap \varphi(A) = \varphi(A_0) \cap B$. But $A_0 \in \underline{\underline{E}}$ and it is well known that the image $\varphi(A_0)$ of a Borel set A_0 in E is analytic and hence universally measurable in F. See, for example, III-T13 and III-24 of [8]. Consequently $\varphi(A_0) \in \underline{\underline{F}}^*$ and since $\varphi(A_1) = \varphi(A_0) \cap B$, one has $\varphi(A_1) \in \underline{\underline{B}}^*$ which proves that ψ is a measurable map from $(B, \underline{\underline{B}}^*)$ to $(A, \underline{\underline{A}})$.

Let μ be a finite measure on $(F, \underline{\underline{F}}^*)$ and choose $B_0 \in \{G \in \underline{\underline{F}}: G \supset B\}$ of minimal μ measure. Let $\tilde{\mu}$ be the trace of μ on $\underline{\underline{B}}^* = \underline{\underline{F}}^*|_B$. See (8.3). Now $\nu = \psi(\mu)$ is a finite measure on $(A, \underline{\underline{A}})$. By (8.5), $\underline{\underline{A}}^* = \underline{\underline{E}}^*|_A$ and so ν may be regarded as measure on $(E, \underline{\underline{E}}^*)$ that is carried by A. Therefore there exists $A_1 \subset A$, $A_1 \in \underline{\underline{E}}$ with $\nu(A_1) = \nu(A)$. Let $B_1 = \varphi(A_1) = \psi^{-1}(A_1) \subset B$. Moreover φ is injective on $A_1 \subset A$, and so by Lusin's theorem (8.7), $B_1 \in \underline{\underline{F}}$. Now $\tilde{\mu}(B_1) = \tilde{\mu}[\psi^{-1}(A_1)] = \nu(A_1) = \nu(A) = \tilde{\mu}(B)$, while $\tilde{\mu}(B) = \mu(B_0)$ and $\tilde{\mu}(B_1) = \mu(B_1 \cap B_0) = \mu(B_1)$. Thus $B_1, B_0 \in \underline{\underline{F}}$ with $B_1 \subset B \subset B_0$ and $\mu(B_1) = \mu(B_0)$. Therefore $B \in \underline{\underline{F}}^*$, proving (14.4).

(14.5) THEOREM. *Let E be a second countable metrizable space. Then the following are equivalent:*

(i) *E is a U-space.*

(ii) *For each metric d on E compatible with the topology of E, E is universally measurable in its d-completion.*

(iii) *There exists a metric d on E compatible with the topology of E such that E is universally measurable in its d-completion.*

PROOF. Since (ii) implies (iii), it suffices to show that (i) implies (ii) and that (iii) implies (i). Suppose E is a U-space. Then there exists a compact metric space $(\hat{E}, d)$ which contains (a homeomorphic image of) E as a universally measurable subspace. Since the d-completion of E is just the closure of E in $\hat{E}$, it follows that E is universally measurable in its d-completion. Consequently (ii) is an immediate consequence of (14.3). Let d be a metric on E compatible with the topology of E and let F be the d-completion of E. Assume that E is universally measurable in F. (We identify E with a dense subset of F.) Since (E, d) is separable its completion F is a Polish space, and consequently F is homeomorphic to a G_δ subset of a compact metric space K. (See Cor. 1, p. 197 of [2].) Let $h: F \to K$ be the homeomorphism. Then $h(F)$ being a G_δ is a Borel subset of K. Since $E \in \underline{\underline{F}}^*$ it follows that $h(E)$ is universally measurable in $h(F)$ and hence in K. But h is a homeomorphism of E onto $h(E) \in \underline{\underline{K}}^*$, and so E is a U-space establishing (14.5).

Recall that if E is a metrizable space, then a finite measure μ on $(E, \underline{\underline{E}})$ is tight if $\mu(E) = \sup\{\mu(K): K \text{ compact}\}$. Moreover if μ is a tight measure on $(E, \underline{\underline{E}})$, then for each $B \in \underline{\underline{E}}$, $\mu(B)$ is the supremum of $\mu(K)$ as K ranges over the compact subsets of B. See [13]. Clearly the same statement is valid for $B \in \underline{\underline{E}}^*$.

(14.6) THEOREM. Let E be a second countable metrizable space. Then E is a U-space if and only if every finite measure on $(E, \underline{\underline{E}})$ is tight.

PROOF. Let E be a U-space and let F be a compact metric space containing E as a universally measurable subspace. Let μ be a finite measure on $(E, \underline{\underline{E}})$. Then we may regard μ as a measure on $(F, \underline{\underline{F}}^*)$ that is carried by E. Since every measure on a compact metric space is tight, there exists an increasing sequence (K_n) of compact subsets of F with each $K_n \subset E$ and $\mu(E) = \sup \mu(K_n)$. But from the definition of the subspace topology each K_n is compact in E and so μ is tight as a measure on $(E, \underline{\underline{E}})$.

Conversely suppose each finite measure on $(E, \underline{\underline{E}})$ is tight. Let d be a metric on E compatible with the topology and let F be the d-completion of E. For simplicity we identify E with a (dense) subset of the complete separable metric space F. In order to show that E is a U-space it suffices by (14.5) to show that E is universally measurable in F. Let μ be a finite measure on $(F, \underline{\underline{F}})$. From (8.4) - see the remark following the proof of (8.5) - one has $\underline{\underline{E}} = \underline{\underline{F}}|_E$ and so we may let $\tilde{\mu}$ be the trace of μ on $\underline{\underline{E}}$ (see (8.3)). Let E_0 be an element of $\underline{\underline{F}}$ containing E of minimal μ measure. Then $\tilde{\mu}(A) = \mu(A_0 \cap E_0)$ whenever $A \in \underline{\underline{E}}$ is of the form $A = A_0 \cap E$ with $A_0 \in \underline{\underline{F}}$. By hypothesis there exists an increasing sequence (K_n) of subsets of E that are compact in E such that $K = \cup K_n \subset E$ and $\tilde{\mu}(K) = \tilde{\mu}(E)$. But it is clear that each K_n is also compact in F and consequently $K \in \underline{\underline{F}}$. Thus $K \subset E \subset E_0$ with $K, E_0 \in \underline{\underline{F}}$ and $\mu(E_0) = \tilde{\mu}(E) = \tilde{\mu}(K) = \mu(K \cap E_0) = \mu(K)$. Hence $E \in \underline{\underline{F}}^*$ proving (14.6).

REMARK. It is not difficult to see that if E is a U-space and μ is a probability on $(E, \underline{\underline{E}})$, then for each sub-$\sigma$-algebra $\underline{\underline{G}} \subset \underline{\underline{E}}$ there exists a regular conditional probability on $(E, \underline{\underline{E}}, \mu)$ given $\underline{\underline{G}}$. See my expository note "On the Construction of Kernels."

We close this section with the following well known fact about metric spaces that will be of importance in the next section.

(14.7) PROPOSITION. Let (E, d) be a metric space and let f be a lower bounded, lower semi-continuous function on E. Then there exists an increasing sequence (f_n) of finite d-uniformly continuous functions on E with $f_n \uparrow f$.

PROOF. By adding a constant to f it suffices to consider the case $f \geq 0$. Also if $f \equiv \infty$ we may take $f_n = n$ and so we may assume that there is an x_0 in E with $f(x_0) < \infty$. Define

$$f_n(x) = \inf_{y \in E} \{f(y) + nd(x, y)\}.$$

Then $0 \leq f_n \leq f$, $f_n(x) \leq f(x_0) + nd(x, x_0) < \infty$ for each x, and $f_n \leq f_{n+1}$. It follows readily from the triangle inequality that

$$|f_n(x) - f_n(y)| \leq nd(x, y)$$

for $x, y \in E$ and so each f_n is d-uniformly continuous. It remains to show that (f_n) increases to f. Fix $y_0 \in E$ and suppose $h < f(y_0)$. Since f is lower semi-continuous there exists a spherical neighborhood $B_\varepsilon(y_0) = \{x: d(x, y_0) < \varepsilon\}$ of y_0 such that $f > h$ on $B_\varepsilon(y_0)$. Choose n such that $n\varepsilon > h$. By considering the cases $y \in B_\varepsilon(y_0)$ and $y \notin B_\varepsilon(y_0)$ separately one sees that $f(y) + nd(y_0, y) > h$ for all $y \in E$. Therefore $f_n(y_0) \geq h$ and so (f_n) increases to f.

The following corollary is an immediate consequence of (14.7).

(14.8) COROLLARY. Let (E, d) be a metric space and $f \in \underset{\sim}{C}_b(E)$. Then there exist monotone sequences (f_n) and (g_n) in $\underset{\sim}{C}_u(E, d)$ with $f_n \uparrow f$ and $g_n \downarrow f$.

15. THE RAY SPACE

The construction of the Ray-Knight compactification $\overline{E}$ in Section 10 and the ensuing development depended on the choice of a compact metric space $\hat{E}$ in which E was embedded as a universally measurable subspace. In view of (14.5) this amounts to a choice of a totally bounded metric d on E compatible with the topology of E. In this section we are going to investigate to what extent our previous results are independent of the choice of d, that is, to what extent they depend only on the topology of E and, of course, the resolvent (U^α).

For the moment we fix a totally bounded metric d on E compatible with the topology of E and let $\hat{E}$ be the d-completion of E. Since E is a U-space it follows from (14.5) that E is universally measurable in $\hat{E}$. One then constructs the Ray-Knight compactification as in Section 10. In particular, one constructs the Ray cone $\underset{\sim}{R}(d)$ and the metric ρ - see (10.1) and (10.4). Recall that $\underset{\sim}{C}_b(E)$ denotes the bounded continuous functions on E. Similarly $\underset{\sim}{C}_b(E, r)$ denotes the bounded Ray continuous functions on E, that is, continuous in the topology induced by ρ. Also $\underset{\sim}{C}_u(d)$, resp. $\underset{\sim}{C}_u(\rho)$, denotes the bounded d-uniformly, resp. ρ-uniformly, continuous functions on E.

The following proposition should be compared with (12.1).

(15.1) PROPOSITION. For each $\alpha > 0$, $U^\alpha \underset{\sim}{C}_b(E) \subset \underset{\sim}{C}_b(E, r)$ and $U^\alpha \underset{\sim}{C}_b(E, r) \subset \underset{\sim}{C}_b(E, r)$.

PROOF. Let $f \in \underset{\sim}{C}_b(E)$. Then by (14.8) there exist monotone sequences (f_n) and (g_n) in $\underset{\sim}{C}_u(d)$ such that $f_n \uparrow f$ and $g_n \downarrow f$. But for each n and $\alpha > 0$, $U^\alpha f_n$ and $U^\alpha g_n$ are in $\underset{\sim}{R}(d) \subset \underset{\sim}{C}_b(E, r)$, and $U^\alpha f_n \uparrow U^\alpha f$ while $U^\alpha g_n \downarrow U^\alpha f$. Consequently $U^\alpha f$ is both lower and upper Ray semi-continuous, and so $U^\alpha f \in \underset{\sim}{C}_b(E, r)$. Similarly starting with the fact that $U^\alpha \underset{\sim}{C}_u(\rho) \subset \underset{\sim}{C}_u(\rho)$ - see (12.1) - it follows that $U^\alpha \underset{\sim}{C}_b(E, r) \subset \underset{\sim}{C}_b(E, r)$.

We are now in a position to show that the Ray topology on E is independent of the choice of the metric d.

(15.2) PROPOSITION. Let d_1 and d_2 be two totally bounded metrics on E compatible with the topology of E. Then the corresponding Ray topologies r_1 and r_2 are the same.

PROOF. Let $\underset{\sim}{R}(d_1)$ and $\underset{\sim}{R}(d_2)$ be the Ray cones constructed in (10.1) from d_1 and d_2 respectively. Since r_1 (resp. r_2) is the weakest topology on E relative to which the elements of $\underset{\sim}{R}(d_1)$ (resp. $\underset{\sim}{R}(d_2)$) are continuous, the desired conclusion will follow in view of the symmetry between d_1 and d_2 once we show $\underset{\sim}{R}(d_1) \subset \underset{\sim}{C}_b(E, r_2)$. By (15.1) for each $\alpha > 0$,

$$U^\alpha \underset{\sim}{C}_u(d_1) \subset U^\alpha \underset{\sim}{C}_b(E) \subset U^\alpha \underset{\sim}{C}_b(E, r_2) \ .$$

But $\underset{\sim}{C}_b^+(E, r_2)$ is a convex cone closed under "$\wedge$" and, by (15.1) again, $U^\alpha \underset{\sim}{C}_b^+(E, r_2) \subset \underset{\sim}{C}_b^+(E, r_2)$ for each $\alpha > 0$. Consequently from the very definition of $\underset{\sim}{R}(d_1)$ one has $\underset{\sim}{R}(d_1) \subset \underset{\sim}{C}_b^+(E, r_2)$, establishing (15.2).

As in previous sections we denote the Ray topology on E by r. We now know that it depends only on the original topology of E and the resolvent (U^α). Of course, E equipped with the Ray topology is a U-space. The next result characterizes the Ray topology without mentioning d.

(15.3) COROLLARY. (i) The Ray topology is the weakest topology τ on E satisfying $U^\alpha \underset{\sim}{C}_b(E) \subset \underset{\sim}{C}_b(E, \tau)$ and $U^\alpha \underset{\sim}{C}_b(E, \tau) \subset \underset{\sim}{C}_b(E, \tau)$ for each $\alpha > 0$.

(ii) Using the notation of the proof of (10.1), let $\underset{\sim}{Q}_0 = \mathcal{U}\underset{\sim}{C}_b^+(E), \ldots,$ $\underset{\sim}{Q}_{n+1} = \Lambda(\underset{\sim}{Q}_n + \mathcal{U}\underset{\sim}{Q}_n), \ldots,$ and $\underset{\sim}{Q} = \cup \underset{\sim}{Q}_n$. Then the Ray topology is the weakest topology on E relative to which the elements of $\underset{\sim}{Q}$ are continuous.

PROOF. By (15.1) the Ray topology has the two properties in (i). If τ is a topology on E having these properties and d is a totally bounded metric on E compatible with the topology of E, then just as in the proof of (15.2) one shows that $\underset{\sim}{R}(d) \subset \underset{\sim}{C}_b^+(E, \tau)$. Consequently the Ray topology is weaker than τ. Coming to (ii) let d be as above. Clearly $\underset{\sim}{R}(d) \subset \underset{\sim}{Q}$ and so if τ is the topology generated by $\underset{\sim}{Q}$, then r is weaker than τ. From (15.1) it is clear that $\underset{\sim}{Q}_0 \subset \underset{\sim}{C}_b^+(E, r)$. Suppose $\underset{\sim}{Q}_n \subset \underset{\sim}{C}_b^+(E, r)$. Then using (15.1) again,

$\underset{\sim}{Q}_n + \mathcal{U}\underset{\sim}{Q}_n \subset \underset{\sim}{C}_b^+(E, r)$ and hence $\underset{\sim}{Q}_{n+1} \subset \underset{\sim}{C}_b^+(E, r)$. As a result $\underset{\sim}{Q} \subset \underset{\sim}{C}_b^+(E, r)$ and so τ is weaker than r, establishing (ii).

REMARK. Of course, $\underset{\sim}{Q}$ is not separable in general.

Let d be a totally bounded metric on E compatible with the topology of E and let $\underset{\sim}{R}(d)$, ρ, $\overline{E}$, $(\overline{U}^\alpha)$, and $(\overline{P}_t)$ be as in Section 10. According to (11.13) the set $N = \{x \in \overline{E}: \overline{P}_0(x, \overline{E} - E) > 0\}$ is useless. Clearly $N \in \underline{\underline{\overline{E}}}^*$ and $N \subset \overline{E} - E$.

(15.4) LEMMA. <u>The set</u> $M = \{x \in \overline{E}: \overline{U}^\alpha(x, \overline{E} - E) > 0\}$ <u>is independent of</u> $\alpha > 0$ <u>and</u> $M \in \underline{\underline{\overline{E}}}^*$. <u>In addition</u>, $M \subset N$ <u>and hence</u> M <u>is useless</u>.

PROOF. For the moment let $M_\alpha = \{x \in \overline{E}: \overline{U}^\alpha(x, \overline{E} - E) > 0\}$. Clearly $M_\alpha \in \underline{\underline{\overline{E}}}^*$. Fix $x \in \overline{E} - M_\alpha$. Then $\overline{U}^\alpha(x, \cdot)$ is carried by E and so

$$\overline{U}^\beta(x, \cdot) = \overline{U}^\alpha(x, \cdot) + (\alpha - \beta) \int_E \overline{U}^\alpha(x, dy)\, \overline{U}^\beta(y, \cdot) .$$

But $\overline{U}^\beta(y, \cdot)$ is carried by E if $y \in E$, and so it follows that $\overline{U}^\beta(x, \cdot)$ is carried by E. This implies that $M_\alpha = M_\beta$ since α and β were arbitrary in the above discussion. If $x \in \overline{E} - N$ and $\alpha > 0$, then since $\overline{U}^\alpha = \overline{P}_0 \overline{U}^\alpha$ one has

$$\overline{U}^\alpha(x, \cdot) = \int_E \overline{P}_0(x, dy)\, \overline{U}^\alpha(y, \cdot)$$

and so $\overline{U}^\alpha(x, \cdot)$ is carried by E. Therefore $M \subset N$, completing the proof of (15.4).

(15.5) DEFINITION. <u>The Ray space of the right semigroup</u> (P_t) <u>is the set</u> $R = \overline{E} - M$ <u>with the subspace topology it inherits from</u> $\overline{E}$.

At first glance it appears that R depends on the metric d, or more precisely the uniformity generated by d, through the space $\overline{E}$. However, we shall show that this is not the case; that is, that R and the restrictions to R of $(\overline{U}^\alpha)$ and $(\overline{P}_t)$ are uniquely determined by the original topology of E and the original resolvent (U^α). By (15.4), $\overline{E} - R = M$ is useless and so in discussing the left limits of X the space R suffices just as well as all of $\overline{E}$. Note that $E \subset R$, that R is a U-space, that E is universally measurable and dense in R, and that the topology R induces on E is the Ray topology. Also R suffices to represent bounded entrance laws for (P_t), but we shall not discuss that here. (See [6].) Consequently the space R has all of the properties of $\bar{E}$ that are relevant to the study of the original right process X. Of course, we must give up the compactness of $\bar{E}$, but as we shall see we gain the fact that R does not depend on the choice of the metric d.

Here is another property of R.

(15.6) PROPOSITION. If $x \in R$, then $\bar{P}_0(x, \cdot)$ is carried by R and $\bar{P}_t(x, \cdot)$ is carried by E for each $t > 0$.

PROOF. If $x \in R$, then since $\bar{U}^\alpha = \bar{P}_0 \bar{U}^\alpha$

$$0 = \bar{U}^\alpha(x, \bar{E} - E) = \int \bar{P}_0(x, dy)\, \bar{U}^\alpha(y, \bar{E} - E) ,$$

and hence $\bar{P}_0(x, \cdot)$ is carried by R. Because $\bar{U}^\alpha(x, \bar{E} - E) = 0$ there exists a sequence (t_n) depending on x of strictly positive numbers decreasing to zero with $\bar{P}_{t_n}(x, \bar{E} - E) = 0$ for each n. If $t > t_n$ for some n, then

$$\bar{P}_t(x, \cdot) = \int_E \bar{P}_{t_n}(x, dy)\, \bar{P}_{t-t_n}(y, \cdot) ,$$

and since $\bar{P}_s(y, \cdot) = P_s(y, \cdot)$ for $y \in E$ and $s > 0$, it follows that $\bar{P}_t(x, \cdot)$ is carried by E for each $t > 0$ and $x \in R$, proving (15.6).

We are now going to show that R is independent of the choice of d. To this end let d_1 and d_2 be two totally bounded metrics on E compatible with the original topology of E. With the obvious notation we are going to show that there exists a homeomorphism of R_2 onto R_1 that leaves E fixed and preserves the semigroup and resolvent. We begin by considering the case in which $d_1 \le d_2$. Thus until further notice d_1 and d_2 are totally bounded metrics on E compatible with the original topology of E and satisfying $d_1 \le d_2$. It is immediate that $\underset{\sim}{C}_u(d_1) \subset \underset{\sim}{C}_u(d_2)$, and hence $\underset{\sim}{R}(d_1) \subset \underset{\sim}{R}(d_2)$. As a result one may choose the metrics ρ_1 and ρ_2 in (10.4) to satisfy $\rho_1 \le \rho_2$. Let $\bar{E}_1$ and $\bar{E}_2$ be the completions of (E, ρ_1) and (E, ρ_2) respectively. For convenience we shall identify E simultaneously with a dense universally measurable subset of $\bar{E}_1$ and $\bar{E}_2$. Since the identity map from (E, ρ_2) to (E, ρ_1) is uniformly continuous there exists a uniformly continuous map φ: $\bar{E}_2 \to \bar{E}_1$ that reduces to the identity on E.

According to (14.1-i), φ is a surjection of $\bar{E}_2$ on $\bar{E}_1$, and with the present identification (14.1-ii) takes the form $\varphi^{-1}(E) = E$. Of course, ρ_1 and ρ_2 both induce the Ray topology on E in light of (15.2).

Let $(\bar{U}_1^\alpha)$ and $(\bar{P}_t^1)$ be the Ray resolvent and semigroup on $\bar{E}_1$ constructed in Section 10, and let $M_1 = \{x \in \bar{E}_1 : \bar{U}_1^\alpha(x, \bar{E}_1 - E) > 0\}$ and $R_1 = \bar{E}_1 - M_1$. The objects $(\bar{U}_2^\alpha)$, $(\bar{P}_t^2)$, M_2, and R_2 are defined similarly relative to $\bar{E}_2$. We come now to the key technical lemma of this development.

(15.7) LEMMA. Using the above notation, for each $x \in \bar{E}_2$ one has $\varphi \bar{U}_2^\alpha(x, \cdot) = \bar{U}_1^\alpha(\varphi(x), \cdot)$ for each $\alpha > 0$, and $\varphi \bar{P}_t^2(x, \cdot) = \bar{P}_t^1(\varphi(x), \cdot)$ for each $t \geq 0$.

PROOF. Of course, $\varphi \bar{U}_2^\alpha(x, \cdot)$ is the measure on $\bar{E}_1$ defined by $A \to \bar{U}_2^\alpha(x, \varphi^{-1}(A))$ for all $A \in \bar{\underline{E}}_1$ (or $\bar{\underline{E}}_1^*$), and $\varphi \bar{P}_t^2(x, \cdot)$ is defined similarly. Suppose first that x is in E. Then $\varphi(x) = x$ and by (11.9)

$$\bar{U}_2^\alpha(x, \cdot) = U^\alpha(x, \cdot) = \bar{U}_1^\alpha(x, \cdot) = \bar{U}_1^\alpha(\varphi(x), \cdot) \quad . \tag{15.8}$$

Since φ is the identity on E and $\varphi^{-1}(E) = E$ it follows that $\varphi\bar{U}_2^\alpha(x, \cdot) = \varphi U^\alpha(x, \cdot) = U^\alpha(x, \cdot)^{(1)}$ and so $\varphi\bar{U}_2^\alpha(x, \cdot) = \bar{U}_1^\alpha(\varphi(x), \cdot)$ for $x \in E$. But $\bar{U}_j^\alpha$ sends $\underset{\sim}{C}(\bar{E}_j)$ into itself, and consequently the map $x \to \bar{U}_j^\alpha(x, \cdot)$ from the compact metric space $\bar{E}_j$ to the bounded measures on $\bar{E}_j$ with the usual weak topology for measures is continuous for $j = 1, 2$. Given $x \in \bar{E}_2$ choose $(x_n) \subset E$ with $x_n \to x$. Since φ is continuous one has $\varphi\bar{U}_2^\alpha(x_n, \cdot) \to \varphi\bar{U}_2^\alpha(x, \cdot)$ and $\bar{U}_1^\alpha(\varphi(x_n), \cdot) \to \bar{U}_1^\alpha(\varphi(x), \cdot)$. Combining these facts yields (i).

For (ii), it follows from (i) that

$$\int_0^\infty e^{-\alpha t} \bar{P}_t^2(x, f \circ \varphi)\, dt = \int_0^\infty e^{-\alpha t} \bar{P}_t^1(\varphi(x), f)\, dt$$

for all $f \in \underset{\sim}{C}(\bar{E}_1)$ and $x \in \bar{E}_2$. Since $f \circ \varphi \in \underset{\sim}{C}(\bar{E}_2)$, both $t \to \bar{P}_t^2(x, f \circ \varphi)$ and $t \to \bar{P}_t^1(\varphi(x), f)$ are right continuous. Therefore the uniqueness theorem for Laplace transforms yields (ii).

(15.9) COROLLARY. $\varphi(R_2) = R_1$ and $\varphi(M_2) = M_1$.

PROOF. If $x \in \bar{E}_2$, then (15.7) and the fact that $\varphi^{-1}(E) = E$ give

---

(1) See the second paragraph on page 108.

$$\bar{U}_1^{\alpha}(\varphi(x), \bar{E}_1 - E) = \bar{U}_2^{\alpha}(x, \varphi^{-1}(\bar{E}_1 - E)) = \bar{U}_2^{\alpha}(x, \bar{E}_2 - E) .$$

Consequently x is in R_2 (resp. M_2) if and only if $\varphi(x)$ is in R_1 (resp. M_1), proving (15.9).

The next proposition is the main result under the present assumption that $d_1 \le d_2$.

(15.10) PROPOSITION. φ <u>is a homeomorphism of</u> R_2 <u>onto</u> R_1 <u>such that</u>:

(i) φ <u>is the identity on</u> E.

(ii) <u>For each</u> $x \in R_2$ <u>and</u> $\alpha > 0$, $\bar{U}_2^{\alpha}(x, \cdot) = \bar{U}_1^{\alpha}(\varphi(x), \cdot)$.

(iii) <u>For each</u> $x \in R_2$, $\bar{P}_t^2(x, \cdot) = \bar{P}_t^1(\varphi(x), \cdot)$ if $t > 0$ <u>and</u> $\varphi\bar{P}_0^2(x, \cdot) = \bar{P}_0^1(\varphi(x), \cdot)$.

PROOF. We begin by showing that φ is a homeomorphism of R_2 onto R_1. By (15.9), φ is a continuous surjection of R_2 onto R_1 and so we must show that φ is injective on R_2 and that φ^{-1} is continuous. We shall first show that φ is injective on R_2. To this end fix x and y in R_2 with $\varphi(x) = \varphi(y)$. Let $g \in b\underline{\underline{E}}^*$. Define f on $\bar{E}_1$ by $f = g$ on E and $f = 0$ on $\bar{E}_1 - E$. Then $f \in b\bar{\underline{\underline{E}}}_1^*$. Since $x, y \in R_2$ the measures $\bar{U}_2^{\alpha}(x, \cdot)$ and $\bar{U}_2^{\alpha}(y, \cdot)$ are carried by E, and so using (15.7)

$$\int_E \bar{U}_2^{\alpha}(x, dz)\, g(z) = \int_E \bar{U}_2^{\alpha}(x, dz)\, f \circ \varphi(z) = \bar{U}_2^{\alpha}(x, f \circ \varphi)$$

$$= \bar{U}_1^{\alpha}(\varphi(x), f) = \bar{U}_1^{\alpha}(\varphi(y), f) = \int_E \bar{U}_2^{\alpha}(y, dz)\, g(z) .$$

Therefore $\bar{U}_2^{\alpha}(x, \cdot) = \bar{U}_2^{\alpha}(y, \cdot)$ for each $\alpha > 0$, and it was shown in the proof of (10.9) that this implies $x = y$. Consequently φ is a continuous bijection of R_2 onto R_1.

In order to show that φ^{-1} from R_1 to R_2 is continuous it suffices to show that φ from R_2 to R_1 is closed. Accordingly we shall show that if A is closed in $\bar{E}_2$, then $\varphi(A \cap R_2)$ is closed in R_1. Now A being closed in $\bar{E}_2$ is compact and so $\varphi(A)$ is compact and hence closed in $\bar{E}_1$. Therefore it suffices to show that $\varphi(A \cap R_2) = \varphi(A) \cap R_1$. It is clear that

$$\varphi(A \cap R_2) \subset \varphi(A) \cap \varphi(R_2) = \varphi(A) \cap R_1 .$$

Conversely if $x \in \varphi(A) \cap R_1$, then there exists $y \in A$ with $x = \varphi(y)$. Now $\varphi(y) = x \in R_1$ and so (15.9) implies that $y \in R_2$. Hence $x \in \varphi(A \cap R_2)$ and therefore $\varphi(A \cap R_2) = \varphi(A) \cap R_1$ completing the proof that φ is a homeomorphism of R_2 onto R_1.

Condition (i) of (15.10) is clear, and (ii) and (iii) of (15.10) are immediate consequences of the definition of R_1 and R_2, (15.6), (15.7), and the easily verified fact that if ν is a measure on R_2 that is carried by E, then $\varphi(\nu) = \nu$. Of course, $\varphi(\nu)$ is a measure on R_1 in general, but if ν is carried by E, then $\varphi(\nu)$ is carried by E and can be identified with ν. This is the meaning of $\varphi(\nu) = \nu$. Since $\bar{P}_0^2(x, \cdot)$ need not be carried by E one can only assert that $\varphi\bar{P}_0^2(x, \cdot) = \bar{P}_0^1(\varphi(x), \cdot)$ for $x \in R_2$. This completes the proof of (15.10).

REMARK. In general φ is <u>not</u> injective on all of $\bar{E}_2$. The reader is invited to furnish an example.

The following theorem is the main result of this section.

(15.11) THEOREM. *Let d_1 and d_2 be two totally bounded metrics on E compatible with the original topology of E. Then there exists a homeomorphism φ of R_2 onto R_1 satisfying (i), (ii), and (iii) of (15.10).*

PROOF. Let $d = d_1 + d_2$. It is easy to check that d is a totally bounded metric on E compatible with the topology. Since $d_1 \le d$ and $d_2 \le d$ we may apply (15.10) to obtain, using the obvious notation, homeomorphisms φ_1 of R onto R_1 and φ_2 of R onto R_2 satisfying the conditions of (15.10). Then $\varphi = \varphi_1 \circ \varphi_2^{-1}$ is a homeomorphism of R_2 onto R_1 and we leave it to the reader to check that φ satisfies conditions (i), (ii), and (iii) of (15.10).

We shall call a homeomorphism of R_2 onto R_1 that satisfies (i), (ii), and (iii) of (15.10) natural. In the general case the natural homeomorphism φ of R_2 onto R_1 is not defined as a map on all of $\bar{E}_2$ as it is when $d_2 \le d_1$. The next result states that this natural homeomorphism preserves the branch points. The notation and assumptions are as in the statement of (15.11).

(15.12) PROPOSITION. *Let B_1 and B_2 be the sets of branch points of $(\bar{P}_t^1)$ and $(\bar{P}_t^2)$ respectively. Then $\varphi(B_2 \cap R_2) = B_1 \cap R_1$ and $\varphi(B_2^c \cap R_2) = B_1^c \cap R_1$.*

PROOF. Recall that $x \in B_j$ if and only if $\bar{P}_0^j(x, \cdot) \neq \varepsilon_x$ and that by (15.6), $\bar{P}_0^j(x, \cdot)$ is carried by R_j if $x \in R_j$ for $j = 1$ or 2. Let $x \in R_2$. If $x \in B_2^c$, then

$$\bar{P}_0^1(\varphi(x), \cdot) = \varphi \bar{P}_0^2(x, \cdot) = \varphi \varepsilon_x = \varepsilon_{\varphi(x)}$$

and so $\varphi(x) \in B_1^c$. Conversely if $\varphi(x) \in B_1^c$ and g is a bounded Borel function on R_2, then $f = g \circ \varphi^{-1}$ is a bounded Borel function on R_1 and

$$\bar{P}_0^2(x, g) = \bar{P}_0^2(x, f \circ \varphi) = \bar{P}_0^1(\varphi(x), f) = f \circ \varphi(x) = g(x) .$$

But $\bar{P}_0^2(x, \cdot)$ is carried by R_2 and so this implies that $x \in B_2^c$. Combining these facts yields (15.12).

REMARK. Recall the definition of the sets N_1 and N_2 above (15.4). An argument similar to the proof of (15.12) shows that $\varphi(N_2 \cap R_2) = N_1 \cap R_1$ and $\varphi(N_2^c \cap R_2) = N_1^c \cap R_1$.

It follows from (15.11) that for each $x \in R$, $\bar{U}^\alpha(x, \cdot)$ and $\bar{P}_t(x, \cdot)$ are uniquely defined independently of the choice of the metric d used in the construction. Moreover for each x in R and $\alpha > 0$, $\bar{U}^\alpha(x, \cdot)$ is carried by E, while according to (15.6), $\bar{P}_t(x, \cdot)$ is carried by E if $t > 0$ and by R if $t = 0$. Since $\underline{\underline{R}}$, the σ-algebra of Borel subsets of R, is the trace of $\underline{\underline{\bar{E}}}$ on R, the same argument used in the proof of (12.1) shows that $\bar{U}^\alpha$ and $\bar{P}_t$ map $b\underline{\underline{R}}$ into $b\underline{\underline{R}}$. Moreover it is now clear that $(\bar{U}^\alpha)$, resp. $(\bar{P}_t)$, is a resolvent, resp. semigroup, of kernels on $(R, \underline{\underline{R}})$.

If $x \in R$, then $\bar{U}^\alpha(x, \cdot)$ is carried by E, and so $f \to \bar{U}^\alpha f$ may be regarded as a map from functions defined on E to functions defined on R. This point of view is adopted in the next result.

(15.13) PROPOSITION. (i) For each $\alpha > 0$, $\bar{U}^\alpha$ maps $\underset{\sim}{C}_b(E)$ and $\underset{\sim}{C}_b(E, r)$ into $\underset{\sim}{C}_b(R)$.

(ii) For each $f \in \underset{\sim}{C}_b(R)$ and $x \in R$, $t \to \bar{P}_t f(x)$ is right continuous on $[0, \infty)$.

PROOF. Let d be a totally bounded metric on E compatible with the topology, and let ρ, $\bar{E}$, and R be defined in terms of d. Statement (i) is now proved by exactly the same argument as that used in the proof of (15.1). For (ii), if $f \in \underset{\sim}{C}_u(R, \rho)$, then there exists a unique $g \in \underset{\sim}{C}(\overline{E})$ with $f = g|_R$, and so for each $t \geq 0$, $\overline{P}_t f = \overline{P}_t g$ on R. Therefore $t \to \overline{P}_t f(x)$ is right continuous on $[0, \infty)$ for each $x \in R$ and $f \in \underset{\sim}{C}_u(R, \rho)$. But if $f \in \underset{\sim}{C}_b(R)$, then there exist monotone sequences (f_n) and (g_n) in $\underset{\sim}{C}_u(R, \rho)$ with $f_n \uparrow f$ and $g_n \downarrow f$. It is easy to see that this implies that $t \to \overline{P}_t f(x)$ is right continuous on $[0, \infty)$ for each $x \in R$, establishing (15.13-ii).

The next proposition shows exactly how the topology of the U-space R is determined by the resolvent $(\overline{U}^\alpha)$ on R. Recall that a sequence of bounded measures (ν_n) on a metrizable space F converges to a bounded measure ν if and only if $\nu_n(f) \to \nu(f)$ for all $f \in \underset{\sim}{C}_b(F)$.

(15.14) PROPOSITION. Let $(x_n) \subset R$ and $x \in R$. Then $x_n \to x$ if and only if for each $\alpha > 0$, $\overline{U}^\alpha(x_n, \cdot) \to \overline{U}^\alpha(x, \cdot)$ as measures on E with the Ray topology.

PROOF. By (15.13-i), $\overline{U}^\alpha$ maps $\underset{\sim}{C}_b(E, r)$ into $\underset{\sim}{C}_b(R)$ and so if $x_n \to x$ in R, then $\overline{U}^\alpha(x_n, \cdot) \to \overline{U}^\alpha(x, \cdot)$ as measures on (E, r). For the converse fix an appropriate metric d and the corresponding ρ and $\overline{E}$. In order to show that $x_n \to x$ it suffices to show that every subsequence of (x_n) contains a further subsequence which converges to x. Changing notation it suffices to show that (x_n) contains a subsequence converging to x whenever $\overline{U}^\alpha(x_n, \cdot) \to \overline{U}^\alpha(x, \cdot)$ for each $\alpha > 0$. But $\overline{E}$ is compact and so (x_n) has a subsequence (x'_n) converging to some point $y \in \overline{E}$. Since $\overline{U}^\alpha$ maps $\underset{\sim}{C}(\overline{E})$ into $\underset{\sim}{C}(\overline{E})$ it follows that $\overline{U}^\alpha(x'_n, \cdot) \to \overline{U}^\alpha(y, \cdot)$ as measures on $\overline{E}$. But for each n, $\overline{U}^\alpha(x'_n, \cdot)$ is carried by E and so is $\overline{U}^\alpha(x, \cdot)$. Since the restriction to E of any function in $\underset{\sim}{C}(\overline{E})$ is in $\underset{\sim}{C}_u(E, \rho) \subset \underset{\sim}{C}_b(E, r)$ it follows from the hypothesis that $\overline{U}^\alpha(x'_n, \cdot) \to \overline{U}^\alpha(x, \cdot)$ as measures on $\overline{E}$. Therefore $\overline{U}^\alpha(x, \cdot) = \overline{U}^\alpha(y, \cdot)$ as measures on $\overline{E}$ for all $\alpha > 0$, and as in the proof of (10.9) this implies that $x = y$. Hence $x'_n \to x$ completing the proof of (15.14).

(15.15) REMARK. An immediate corollary to (15.14) is the fact that given $(x_n) \subset E$ and $x \in E$, then $x_n \to x$ in the Ray topology of E if and only if for each $\alpha > 0$, $U^\alpha(x_n, \cdot) \to U^\alpha(x, \cdot)$ as measures on E with the Ray topology.

We know that E equipped with the Ray topology is a U-space. Therefore if we regard X as an (E, r) valued process, then X is a right process with resolvent (U^α). It is natural to ask what happens if we apply the Ray-Knight procedure again to X as an (E, r) process. We shall show that we get nothing new by this procedure, at least as far as the Ray space is concerned.

To this end let d be a totally bounded metric on E compatible with the original topology of E. Starting from d construct ρ, $\overline{E}$, R, and the Ray topology r as before. Next let δ be a totally bounded metric on E that is compatible with the Ray topology of E. As in Section 10 we construct the Ray cone $\underset{\sim}{R}(\delta)$ relative to (E, δ). To be explicit using the notation of the proof of (10.1), we have

$$\underset{\sim}{R}_0(\delta) = \mathcal{U}\underset{\sim}{C}_u^+(E, \delta), \ldots, \underset{\sim}{R}_{n+1}(\delta) = \Lambda(\underset{\sim}{R}_n(\delta) + \mathcal{U}\underset{\sim}{R}_n(\delta)) ,$$

and $\underset{\sim}{R}(\delta) = \cup \underset{\sim}{R}_n(\delta)$. From $\underset{\sim}{R}(\delta)$ we construct a metric $\rho^\#$ as in (10.4) and we let $\overline{E}^\#$ be the (compact) completion of $(E, \rho^\#)$. We let $(\overline{U}_\#^\alpha)$ denote the corresponding Ray resolvent on $\overline{E}^\#$ and $R^\# = \overline{E}^\# - M^\#$ the corresponding Ray space. Finally $r^\#$ denotes the topology induced on E by $\rho^\#$, or equivalently the subspace topology E inherits from $\overline{E}^\#$.

(15.16) PROPOSITION. <u>The topologies r and $r^\#$ on E are the same.</u>

PROOF. In the present situation (15.1) implies that $U^\alpha \underset{\sim}{C}_b(E, r) \subset \underset{\sim}{C}_b(E, r^\#)$ and that $U^\alpha \underset{\sim}{C}_b(E, r^\#) \subset \underset{\sim}{C}_b(E, r^\#)$ for each $\alpha > 0$. Of course, $U^\alpha \underset{\sim}{C}_b(E, r) \subset \underset{\sim}{C}_b(E, r)$ by (15.1). Since $\underset{\sim}{C}_u(E, \delta) \subset \underset{\sim}{C}_b(E, r)$, it follows that $\underset{\sim}{R}(\delta) \subset \underset{\sim}{C}_b(E, r)$. This in turn implies that $r^\#$ is weaker than r. Because both r and $r^\#$ are metrizable in order to complete the proof of (15.16) it suffices to show that if $x_n \to x$ in $r^\#$, then $x_n \to x$ in r whenever $(x_n) \subset E$ and $x \in E$. Let $f \in \underset{\sim}{C}_b(E, r)$. Then $U^\alpha f \in \underset{\sim}{C}_b(E, r^\#)$ and so if $x_n \to x$ in $r^\#$, then $U^\alpha f(x_n) \to U^\alpha f(x)$. But this is just the statement that $U^\alpha(x_n, \cdot) \to U^\alpha(x, \cdot)$ as measures on (E, r), and hence by (15.15), $x_n \to x$ in r. This establishes (15.16).

We are now going to show that R and $R^\#$ are naturally isomorphic. The notation is that introduced above (15.16). We identify E with a universally measurable subspace of R and $R^\#$ simultaneously. Thus $E \subset R \subset \overline{E}$ and $E \subset R^\# \subset \overline{E}^\#$. Also recall that $\overline{U}^\alpha(x, \cdot)$, resp. $\overline{U}_\#^\alpha(x, \cdot)$, is carried by E for $x \in R$, resp. for $x \in R^\#$, and that a similar statement holds for $\overline{P}_t(x, \cdot)$ and $\overline{P}_t^\#(x, \cdot)$ if $t > 0$.

(15.17) THEOREM. There exists a homeomorphism ψ of R onto $R^{\#}$ satisfying:

(i) ψ is the identity on E.

(ii) $\bar{U}^{\alpha}(x, \cdot) = \bar{U}^{\alpha}_{\#}(\psi(x), \cdot)$ for each $x \in R$ and $\alpha > 0$.

(iii) For each $x \in R$, $\bar{P}_t(x, \cdot) = \bar{P}^{\#}_t(\psi(x), \cdot)$ if $t > 0$ and $\psi \bar{P}_0(x, \cdot) = \bar{P}^{\#}_0(\psi(x), \cdot)$.

PROOF. Since E is dense in R, given $x \in R$ there exists $(x_n) \subset E$ with $x_n \to x$ in R. Hence by (15.14), $U^{\alpha}(x_n, \cdot) \to \bar{U}^{\alpha}(x, \cdot)$ as measures on (E, r) for each $\alpha > 0$. But $(x_n) \subset E \subset R^{\#} \subset \bar{E}^{\#}$ and so (x_n) has a subsequence, call it (x_n) again, converging to some point $z \in \bar{E}^{\#}$. Let $f \in \underset{\sim}{C}(\bar{E}^{\#})$. Then

$$U^{\alpha}(f|_E)(x_n) = \bar{U}^{\alpha}_{\#} f(x_n) \to \bar{U}^{\alpha}_{\#} f(z) .$$

Now $f|_E$ is $r^{\#}$ and hence r continuous, and so $U^{\alpha}(f|_E)(x_n) \to \bar{U}^{\alpha}(f|_E)(x)$. Hence for all $f \in \underset{\sim}{C}(\bar{E}^{\#})$ and $\alpha > 0$

$$\bar{U}^{\alpha}(f|_E)(x) = \bar{U}^{\alpha}_{\#} f(z) . \tag{15.18}$$

Since both expressions in (15.18) are measures on $\bar{E}^{\#}$, it follows that $\bar{U}^{\alpha}_{\#}(z, \cdot)$ is carried by E, that is, $z \in R^{\#}$. Moreover z depends only on x and not on the particular sequence (x_n) used in its construction. To see this suppose that a second sequence leads to a $z' \in R^{\#}$. Then because of (15.18) one has $\bar{U}^{\alpha}_{\#}(z, \cdot) = \bar{U}^{\alpha}_{\#}(z', \cdot)$ for all $\alpha > 0$, and this implies $z = z'$. (See the proof of (10.9).) We now define $\psi(x) = z$. It is obvious that $\psi(x) = x$ if $x \in E$, and from the construction that $\bar{U}^{\alpha}(x, \cdot) = \bar{U}^{\alpha}_{\#}(\psi(x), \cdot)$ for each $\alpha > 0$ and $x \in R$. Since $(\bar{U}^{\alpha})$ and $(\bar{U}^{\alpha}_{\#})$ separate the points of R and $R^{\#}$ respectively and since the same construction is valid if we begin with a point $z \in R^{\#}$, it follows that ψ is a bijection of R onto $R^{\#}$. Using (15.14) it is immediate that ψ is a homeomorphism. In light of (15.13-ii), statement (iii) follows from (ii) just as in the proof of (15.7). This establishes (15.17).

Theorem 15.17 gives a precise meaning to the statement that if one applies the Ray-Knight procedure to E equipped with the Ray topology one obtains nothing new. It is possible to go further and characterize the Ray space R up to a useless set. However, we shall not pursue this here. We refer the interested reader to [6].

BIBLIOGRAPHY

1. R. M. BLUMENTHAL and R. K. GETOOR, "Markov Processes and Potential Theory." Academic Press. New York. (1968).

2. N. BOURBAKI, "General Topology, Part 2." Hermann. Paris. (1966).

3. C. DELLACHERIE, "Capacités et Processus Stochastiques." Springer-Verlag. Heidelberg. (1972).

4. W. FELLER, "An Introduction to Probability Theory and its Applications." Vol. 2, Second Ed. Wiley. New York. (1971).

5. R. K. GETOOR and M. J. SHARPE, "Balayage and multiplicative functionals." Zeit. für Wahrscheinlichkeitstheorie. 28, 139-164 (1974).

6. R. K. GETOOR and M. J. SHARPE, "The Ray space of a right processes." To appear Ann. Instit. Fourier. Grenoble.

7. F. KNIGHT, "Note on regularization of Markov processes." Ill. Journ. Math. 9, 548-552 (1965).

8. P. A. MEYER, "Probability and Potentials." Ginn. Boston. (1966).

9. P. A. MEYER, "Processus de Markov." Lecture Notes in Math. 26, Springer-Verlag. Heidelberg (1967).

10. P. A. MEYER, "Processus de Markov: La Frontière de Martin." Lecture Notes in Math. 77. Springer-Verlag. Heidelberg (1968).

11. P. A. MEYER, "Balayage pour les processus de Markov continus à droite, d'après Shih Chung Tuo." Lecture Notes in Math. 191, 270-274. Springer-Verlag. Heidelberg (1971).

12. P. A. MEYER, "Remarque sur les hypothèses droites." Lecture Notes in Math. 321, 205-209. Springer-Verlag. Heidelberg (1973).

13. K. P. PARTHASARATHY, "Probability Measures on Metric Spaces." Academic Press. New York (1967).

14. D. B. RAY, "Resolvents, transition functions, and strongly Markovian processes." Ann. Math. 70, 43-72 (1959).

15. C. T. SHIH, "On extending potential theory to all strong Markov processes." Ann. Instit. Fourier 20, 303-315 (1970).

16. J. B. WALSH and P. A. MEYER, "Quelques applications des rèsolvantes de Ray." Invent. Math. 14, 143-166 (1971).

ADDED NOTE. The following reference contains a simpler approach to the basic result, Theorem 7.6, characterizing previsible and totally inaccessible stopping times for a Ray process. Unfortunately, it came to our attention too late for us to make use of it in Section 7.

K. L. CHUNG and J. B. WALSH, "Meyer's theorem on predictability." Zeit. für Wahrscheinlichkeitstheorie. 29, 253-256 (1974).

INDEX OF NOTATION

SUBJECT INDEX

Vol. 277: Séminaire Banach. Edité par C. Houzel. VII, 229 pages. 1972. DM 22,-

Vol. 278: H. Jacquet, Automorphic Forms on GL(2) Part II. XIII, 142 pages. 1972. DM 18,-

Vol. 279: R. Bott, S. Gitler and I. M. James, Lectures on Algebraic and Differential Topology. V, 174 pages. 1972. DM 20,-

Vol. 280: Conference on the Theory of Ordinary and Partial Differential Equations. Edited by W. N. Everitt and B. D. Sleeman. XV, 367 pages. 1972. DM 29,-

Vol. 281: Coherence in Categories. Edited by S. Mac Lane. VII, 235 pages. 1972. DM 22,-

Vol. 282: W. Klingenberg und P. Flaschel, Riemannsche Hilbertmannigfaltigkeiten. Periodische Geodätische. VII, 211 Seiten. 1972. DM 22,-

Vol. 283: L. Illusie, Complexe Cotangent et Déformations II. VII, 304 pages. 1972. DM 27,-

Vol. 284: P. A. Meyer, Martingales and Stochastic Integrals I. VI, 89 pages. 1972. DM 18,-

Vol. 285: P. de la Harpe, Classical Banach-Lie Algebras and Banach-Lie Groups of Operators in Hilbert Space. III, 160 pages. 1972. DM 18,-

Vol. 286: S. Murakami, On Automorphisms of Siegel Domains. V, 95 pages. 1972. DM 18,-

Vol. 287: Hyperfunctions and Pseudo-Differential Equations. Edited by H. Komatsu. VII, 529 pages. 1973. DM 40,-

Vol. 288: Groupes de Monodromie en Géométrie Algébrique. (SGA 7 I). Dirigé par A. Grothendieck. IX, 523 pages. 1972. DM 55,-

Vol. 289: B. Fuglede, Finely Harmonic Functions. III, 188. 1972. DM 20,-

Vol. 290: D. B. Zagier, Equivariant Pontrjagin Classes and Applications to Orbit Spaces. IX, 130 pages. 1972. DM 18,-

Vol. 291: P. Orlik, Seifert Manifolds. VIII, 155 pages. 1972. DM 18,-

Vol. 292: W. D. Wallis, A. P. Street and J. S. Wallis, Combinatorics: Room Squares, Sum-Free Sets, Hadamard Matrices. V, 508 pages. 1972. DM 55,-

Vol. 293: R. A. DeVore, The Approximation of Continuous Functions by Positive Linear Operators. VIII, 289 pages. 1972. DM 27,-

Vol. 294: Stability of Stochastic Dynamical Systems. Edited by R. F. Curtain. IX, 332 pages. 1972. DM 29,-

Vol. 295: C. Dellacherie, Ensembles Analytiques Capacités. Mesures de Hausdorff. XII, 123 pages. 1972. DM 18,-

Vol. 296: Probability and Information Theory II. Edited by M. Behara, K. Krickeberg and J. Wolfowitz. V, 223 pages. 1973. DM 22,-

Vol. 297: J. Garnett, Analytic Capacity and Measure. IV, 138 pages. 1972. DM 18,-

Vol. 298: Proceedings of the Second Conference on Compact Transformation Groups. Part 1. XIII, 453 pages. 1972. DM 35,-

Vol. 299: Proceedings of the Second Conference on Compact Transformation Groups. Part 2. XIV, 327 pages. 1972. DM 29,-

Vol. 300: P. Eymard, Moyennes Invariantes et Représentations Unitaires. II, 113 pages. 1972. DM 18,-

Vol. 301: F. Pittnauer, Vorlesungen über asymptotische Reihen. VI, 186 Seiten. 1972. DM 18,-

Vol. 302: M. Demazure, Lectures on p-Divisible Groups. V, 98 pages. 1972. DM 18,-

Vol. 303: Graph Theory and Applications. Edited by Y. Alavi, D. R. Lick and A. T. White. IX, 329 pages. 1972. DM 26,-

Vol. 304: A. K. Bousfield and D. M. Kan, Homotopy Limits, Completions and Localizations. V, 348 pages. 1972. DM 29,-

Vol. 305: Théorie des Topos et Cohomologie Etale des Schémas. Tome 3. (SGA 4). Dirigé par M. Artin, A. Grothendieck et J. L. Verdier. VI, 640 pages. 1973. DM 55,-

Vol. 306: H. Luckhardt, Extensional Gödel Functional Interpretation. VI, 161 pages. 1973. DM 20,-

Vol. 307: J. L. Bretagnolle, S. D. Chatterji et P.-A. Meyer, Ecole d'été de Probabilités: Processus Stochastiques. VI, 198 pages. 1973. DM 22,-

Vol. 308: D. Knutson, λ-Rings and the Representation Theory of the Symmetric Group. IV, 203 pages. 1973. DM 22,-

Vol. 309: D. H. Sattinger, Topics in Stability and Bifurcation Theory. VI, 190 pages. 1973. DM 20,-

Vol. 310: B. Iversen, Generic Local Structure of the Morphisms in Commutative Algebra. IV, 108 pages. 1973. DM 18,-

Vol. 311: Conference on Commutative Algebra. Edited by J. W. Brewer and E. A. Rutter. VII, 251 pages. 1973. DM 24,-

Vol. 312: Symposium on Ordinary Differential Equations. Edited by W. A. Harris, Jr. and Y. Sibuya. VIII, 204 pages. 1973. DM 22,-

Vol. 313: K. Jörgens and J. Weidmann, Spectral Properties of Hamiltonian Operators. III, 140 pages. 1973. DM 18,-

Vol. 314: M. Deuring, Lectures on the Theory of Algebraic Functions of One Variable. VI, 151 pages. 1973. DM 18,-

Vol. 315: K. Bichteler, Integration Theory (with Special Attention to Vector Measures). VI, 357 pages. 1973. DM 29,-

Vol. 316: Symposium on Non-Well-Posed Problems and Logarithmic Convexity. Edited by R. J. Knops. V, 176 pages. 1973. DM 20,-

Vol. 317: Séminaire Bourbaki - vol. 1971/72. Exposés 400-417. IV, 361 pages. 1973. DM 29,-

Vol. 318: Recent Advances in Topological Dynamics. Edited by A. Beck. VIII, 285 pages. 1973. DM 27,-

Vol. 319: Conference on Group Theory. Edited by R. W. Gatterdam and K. W. Weston. V, 188 pages. 1973. DM 20,-

Vol. 320: Modular Functions of One Variable I. Edited by W. Kuyk. V, 195 pages. 1973. DM 20,-

Vol. 321: Séminaire de Probabilités VII. Edité par P. A. Meyer. VI, 322 pages. 1973. DM 29,-

Vol. 322: Nonlinear Problems in the Physical Sciences and Biology. Edited by I. Stakgold, D. D. Joseph and D. H. Sattinger. VIII, 357 pages. 1973. DM 29,-

Vol. 323: J. L. Lions, Perturbations Singulières dans les Problèmes aux Limites et en Contrôle Optimal. XII, 645 pages. 1973. DM 46,-

Vol. 324: K. Kreith, Oscillation Theory. VI, 109 pages. 1973. DM 18,-

Vol. 325: C.-C. Chou, La Transformation de Fourier Complexe et L'Equation de Convolution. IX, 137 pages. 1973. DM 18,-

Vol. 326: A. Robert, Elliptic Curves. VIII, 264 pages. 1973. DM 24,-

Vol. 327: E. Matlis, One-Dimensional Cohen-Macaulay Rings. XII, 157 pages. 1973. DM 20,-

Vol. 328: J. R. Büchi and D. Siefkes, The Monadic Second Order Theory of All Countable Ordinals. VI, 217 pages. 1973. DM 22,-

Vol. 329: W. Trebels, Multipliers for (C, α)-Bounded Fourier Expansions in Banach Spaces and Approximation Theory. VII, 103 pages. 1973. DM 18,-

Vol. 330: Proceedings of the Second Japan-USSR Symposium on Probability Theory. Edited by G. Maruyama and Yu. V. Prokhorov. VI, 550 pages. 1973. DM 40,-

Vol. 331: Summer School on Topological Vector Spaces. Edited by L. Waelbroeck. VI, 226 pages. 1973. DM 22,-

Vol. 332: Séminaire Pierre Lelong (Analyse) Année 1971-1972. V, 131 pages. 1973. DM 18,-

Vol. 333: Numerische, insbesondere approximationstheoretische Behandlung von Funktionalgleichungen. Herausgegeben von R. Ansorge und W. Törnig. VI, 296 Seiten. 1973. DM 27,-

Vol. 334: F. Schweiger, The Metrical Theory of Jacobi-Perron Algorithm. V, 111 pages. 1973. DM 18,-

Vol. 335: H. Huck, R. Roitzsch, U. Simon, W. Vortisch, R. Walden, B. Wegner und W. Wendland, Beweismethoden der Differentialgeometrie im Großen. IX, 159 Seiten. 1973. DM 20,-

Vol. 336: L'Analyse Harmonique dans le Domaine Complexe. Edité par E. J. Akutowicz. VIII, 169 pages. 1973. DM 20,-

Vol. 337: Cambridge Summer School in Mathematical Logic. Edited by A. R. D. Mathias and H. Rogers. IX, 660 pages. 1973. DM 46,-

Vol. 338: J. Lindenstrauss and L. Tzafriri, Classical Banach Spaces. IX, 243 pages. 1973. DM 24,-

Vol. 339: G. Kempf, F. Knudsen, D. Mumford and B. Saint-Donat, Toroidal Embeddings I. VIII, 209 pages. 1973. DM 22,-

Vol. 340: Groupes de Monodromie en Géométrie Algébrique. (SGA 7 II). Par P. Deligne et N. Katz. X, 438 pages. 1973. DM 44,-

Vol. 341: Algebraic K-Theory I, Higher K-Theories. Edited by H. Bass. XV, 335 pages. 1973. DM 29,-

Vol. 342: Algebraic K-Theory II, "Classical" Algebraic K-Theory, and Connections with Arithmetic. Edited by H. Bass. XV, 527 pages. 1973. DM 40,-

Vol. 343: Algebraic K-Theory III, Hermitian K-Theory and Geometric Applications. Edited by H. Bass. XV, 572 pages. 1973. DM 40,-

Vol. 344: A. S. Troelstra (Editor), Metamathematical Investigation of Intuitionistic Arithmetic and Analysis. XVII, 485 pages. 1973. DM 38,-

Vol. 345: Proceedings of a Conference on Operator Theory. Edited by P. A. Fillmore. VI, 228 pages. 1973. DM 22,-

Vol. 346: Fučík et al., Spectral Analysis of Nonlinear Operators. II, 287 pages. 1973. DM 26,-

Vol. 347: J. M. Boardman and R. M. Vogt, Homotopy Invariant Algebraic Structures on Topological Spaces. X, 257 pages. 1973. DM 24,-

Vol. 348: A. M. Mathai and R. K. Saxena, Generalized Hypergeometric Functions with Applications in Statistics and Physical Sciences. VII, 314 pages. 1973. DM 26,-

Vol. 349: Modular Functions of One Variable II. Edited by W. Kuyk and P. Deligne. V, 598 pages. 1973. DM 38,-

Vol. 350: Modular Functions of One Variable III. Edited by W. Kuyk and J.-P. Serre. V, 350 pages. 1973. DM 26,-

Vol. 351: H. Tachikawa, Quasi-Frobenius Rings and Generalizations. XI, 172 pages. 1973. DM 20,-

Vol. 352: J. D. Fay, Theta Functions on Riemann Surfaces. V, 137 pages. 1973. DM 18,-

Vol. 353: Proceedings of the Conference on Orders, Group Rings and Related Topics. Organized by J. S. Hsia, M. L. Madan and T. G. Ralley. X, 224 pages. 1973. DM 22,-

Vol. 354: K. J. Devlin, Aspects of Constructibility. XII, 240 pages. 1973. DM 24,-

Vol. 355: M. Sion, A Theory of Semigroup Valued Measures. V, 140 pages. 1973. DM 18,-

Vol. 356: W. L. J. van der Kallen, Infinitesimally Central-Extensions of Chevalley Groups. VII, 147 pages. 1973. DM 18,-

Vol. 357: W. Borho, P. Gabriel und R. Rentschler, Primideale in Einhüllenden auflösbarer Lie-Algebren. V, 182 Seiten. 1973. DM 20,-

Vol. 358: F. L. Williams, Tensor Products of Principal Series Representations. VI, 132 pages. 1973. DM 18,-

Vol. 359: U. Stammbach, Homology in Group Theory. VIII, 183 pages. 1973. DM 20,-

Vol. 360: W. J. Padgett and R. L. Taylor, Laws of Large Numbers for Normed Linear Spaces and Certain Fréchet Spaces. VI, 111 pages. 1973. DM 18,-

Vol. 361: J. W. Schutz, Foundations of Special Relativity: Kinematic Axioms for Minkowski Space Time. XX, 314 pages. 1973. DM 26,-

Vol. 362: Proceedings of the Conference on Numerical Solution of Ordinary Differential Equations. Edited by D. Bettis. VIII, 490 pages. 1974. DM 34,-

Vol. 363: Conference on the Numerical Solution of Differential Equations. Edited by G. A. Watson. IX, 221 pages. 1974. DM 20,-

Vol. 364: Proceedings on Infinite Dimensional Holomorphy. Edited by T. L. Hayden and T. J. Suffridge. VII, 212 pages. 1974. DM 20,-

Vol. 365: R. P. Gilbert, Constructive Methods for Elliptic Equations. VII, 397 pages. 1974. DM 26,-

Vol. 366: R. Steinberg, Conjugacy Classes in Algebraic Groups (Notes by V. V. Deodhar). VI, 159 pages. 1974. DM 18,-

Vol. 367: K. Langmann und W. Lütkebohmert, Cousinverteilungen und Fortsetzungssätze. VI, 151 Seiten. 1974. DM 16,-

Vol. 368: R. J. Milgram, Unstable Homotopy from the Stable Point of View. V, 109 pages. 1974. DM 16,-

Vol. 369: Victoria Symposium on Nonstandard Analysis. Edited by A. Hurd and P. Loeb. XVIII, 339 pages. 1974. DM 26,-

Vol. 370: B. Mazur and W. Messing, Universal Extensions and One Dimensional Crystalline Cohomology. VII, 134 pages. 1974. DM 16,-

Vol. 371: V. Poenaru, Analyse Différentielle. V, 228 pages. 1974. DM 20,-

Vol. 372: Proceedings of the Second International Conference on the Theory of Groups 1973. Edited by M. F. Newman. VII, 740 pages. 1974. DM 48,-

Vol. 373: A. E. R. Woodcock and T. Poston, A Geometrical Study of the Elementary Catastrophes. V, 257 pages. 1974. DM 22,-

Vol. 374: S. Yamamuro, Differential Calculus in Topological Linear Spaces. IV, 179 pages. 1974. DM 18,-

Vol. 375: Topology Conference 1973. Edited by R. F. Dickman Jr. and P. Fletcher. X, 283 pages. 1974. DM 24,-

Vol. 376: D. B. Osteyee and I. J. Good, Information, Weight of Evidence, the Singularity between Probability Measures and Signal Detection. XI, 156 pages. 1974. DM 16,-

Vol. 377: A. M. Fink, Almost Periodic Differential Equations. VIII, 336 pages. 1974. DM 26,-

Vol. 378: TOPO 72 – General Topology and its Applications. Proceedings 1972. Edited by R. Alò, R. W. Heath and J. Nagata. XIV, 651 pages. 1974. DM 50,-

Vol. 379: A. Badrikian et S. Chevet, Mesures Cylindriques, Espaces de Wiener et Fonctions Aléatoires Gaussiennes. X, 383 pages. 1974. DM 32,-

Vol. 380: M. Petrich, Rings and Semigroups. VIII, 182 pages. 1974. DM 18,-

Vol. 381: Séminaire de Probabilités VIII. Edité par P. A. Meyer. IX, 354 pages. 1974. DM 32,-

Vol. 382: J. H. van Lint, Combinatorial Theory Seminar Eindhoven University of Technology. VI, 131 pages. 1974. DM 18,-

Vol. 383: Séminaire Bourbaki – vol. 1972/73. Exposés 418-435 IV, 334 pages. 1974. DM 30,-

Vol. 384: Functional Analysis and Applications, Proceedings 1972. Edited by L. Nachbin. V, 270 pages. 1974. DM 22,-

Vol. 385: J. Douglas Jr. and T. Dupont, Collocation Methods for Parabolic Equations in a Single Space Variable (Based on C^1-Piecewise-Polynomial Spaces). V, 147 pages. 1974. DM 16,-

Vol. 386: J. Tits, Buildings of Spherical Type and Finite BN-Pairs. IX, 299 pages. 1974. DM 24,-

Vol. 387: C. P. Bruter, Eléments de la Théorie des Matroïdes. V, 138 pages. 1974. DM 18,-

Vol. 388: R. L. Lipsman, Group Representations. X, 166 pages. 1974. DM 20,-

Vol. 389: M.-A. Knus et M. Ojanguren, Théorie de la Descente et Algèbres d' Azumaya. IV, 163 pages. 1974. DM 20,-

Vol. 390: P. A. Meyer, P. Priouret et F. Spitzer, Ecole d'Eté de Probabilités de Saint-Flour III – 1973. Edité par A. Badrikian et P.-L. Hennequin. VIII, 189 pages. 1974. DM 20,-

Vol. 391: J. Gray, Formal Category Theory: Adjointness for 2-Categories. XII, 282 pages. 1974. DM 24,-

Vol. 392: Géométrie Différentielle, Colloque, Santiago de Compostela, Espagne 1972. Edité par E. Vidal. VI, 225 pages. 1974. DM 20,-

Vol. 393: G. Wassermann, Stability of Unfoldings. IX, 164 pages. 1974. DM 20,-

Vol. 394: W. M. Patterson 3rd. Iterative Methods for the Solution of a Linear Operator Equation in Hilbert Space – A Survey. III, 183 pages. 1974. DM 20,-

Vol. 395: Numerische Behandlung nichtlinearer Integrodifferential- und Differentialgleichungen. Tagung 1973. Herausgegeben von R. Ansorge und W. Törnig. VII, 313 Seiten. 1974. DM 28,-

Vol. 396: K. H. Hofmann, M. Mislove and A. Stralka, The Pontryagin Duality of Compact O-Dimensional Semilattices and its Applications. XVI, 122 pages. 1974. DM 18,-

Vol. 397: T. Yamada, The Schur Subgroup of the Brauer Group. V, 159 pages. 1974. DM 18,-

Vol. 398: Théories de l'Information, Actes des Rencontres de Marseille-Luminy, 1973. Edité par J. Kampé de Fériet et C. Picard. XII, 201 pages. 1974. DM 23,-